# Aging-friendly Service System Design

Wen Tang

# Aging-friendly Service System Design

华中科技大学出版社
Huazhong University of Science and Technology Press

Springer

Wen Tang
Design, Academy of Fine Arts
Central China Normal University
Wuhan, Hubei, China

ISBN 978-981-95-2053-4    ISBN 978-981-95-2054-1 (eBook)
https://doi.org/10.1007/978-981-95-2054-1
This work was supported by Hubei Provincial Social Science Fund (HBSKJJ20243412).

Jointly published with Huazhong University of Science and Technology Press
The print edition is not for sale in China (Mainland). Customers from China (Mainland) please order the print book from: Huazhong University of Science and Technology Press
Originally published by Huazhong University of Science and Technology Press



This Springer imprint is published by the registered company Springer Nature Singapore Pte Ltd.
The registered company address is: 152 Beach Road, #21-01/04 Gateway East, Singapore 189721, Singapore

# Preface

In the mighty tide of the new era, population aging has emerged as a significant challenge confronting the global community. The report of the 20th National Congress of the Communist Party of China profoundly articulates the national strategy to actively address population aging, demonstrating the state's elevated prioritization and long-term planning for elderly care initiatives. Against this backdrop, age-friendly design now carries unprecedented historical mission and social responsibility. The publication of this book, Age-Friendly Service System Design, constitutes a proactive response and in-depth exploration of this epochal imperative. It aims to integrate socialist core values into design practices, vigorously promote the traditional virtue of respecting and honoring the elderly, and cultivate design professionals imbued with both social responsibility and humanistic care.

The content system of this book is meticulously structured around core issues of aging societies. Commencing with a penetrating analysis of global trends and challenges in population aging, it comprehensively outlines the macro-context and profound significance of age-friendly service system design. Building upon this foundation, the book systematically elaborates on the definition, importance, policy frameworks, legal regulations, and comprehensive considerations of data security and privacy protection in age-friendly services, establishing a holistic knowledge framework. Subsequently, it delves into the physiological and psychological characteristics of older adults, conducting multidimensional analyses of physical functions, cognitive-perceptual capacities, psychological needs, and evolving social roles. These explorations provide theoretical underpinnings for design practices while revealing the unique needs and challenges of elderly populations, emphasizing the centrality of human-centered care and differentiated design in service systems.

When examining core principles of age-friendly service system design, the book thoroughly explores multiple perspectives including universal design, barrier-free design, safety and usability, as well as respect and inclusivity. These discussions profoundly reflect humanistic concerns and ethical considerations in design, offering clear guidance for decision-making. Furthermore, through introducing scientific methodologies and tools, the book guides readers to deeply understand elderly users' needs and preferences, providing precise orientation for design implementation. Concurrently, it addresses critical elements in interface design and interactive experience optimization—including visual element design, operational process simplification, and predictive design—aiming to holistically enhance user experience and satisfaction for elderly populations.

The rapid advancement of intelligent technologies presents boundless possibilities for innovating age-friendly service systems. Keeping pace with contemporary developments, this book extensively explores the innovative integration of cutting-edge technologies such as IoT, artificial intelligence, and smart home systems in elderly care services. These technological applications not only deliver more convenient and efficient service experiences for older adults but also open vast frontiers for intelligent and personalized development of age-friendly service systems.

The compilation of this book synthesizes expertise from numerous distinguished scholars and practitioners in gerontology, design studies, and information technology. Leveraging rich practical experience and acute academic insights, they have collectively infused the latest research achievements and practical wisdom into this work, presenting readers with a scholarly yet pragmatic masterpiece. We hereby extend heartfelt gratitude and highest respect to all contributors who dedicated efforts to this publication.

Looking forward, we firmly believe that Age-Friendly Service System Design will become a crucial driver in advancing elderly care initiatives, infusing wisdom and vitality into the construction of an elderly-friendly society. Simultaneously, we anticipate more visionaries joining this cause to collaboratively contribute to the well-being of elderly populations and the harmonious development of society.

Wuhan, China Wen Tang

# Contents

# Overview of Aging Societies and Age-Friendly Services

# 1

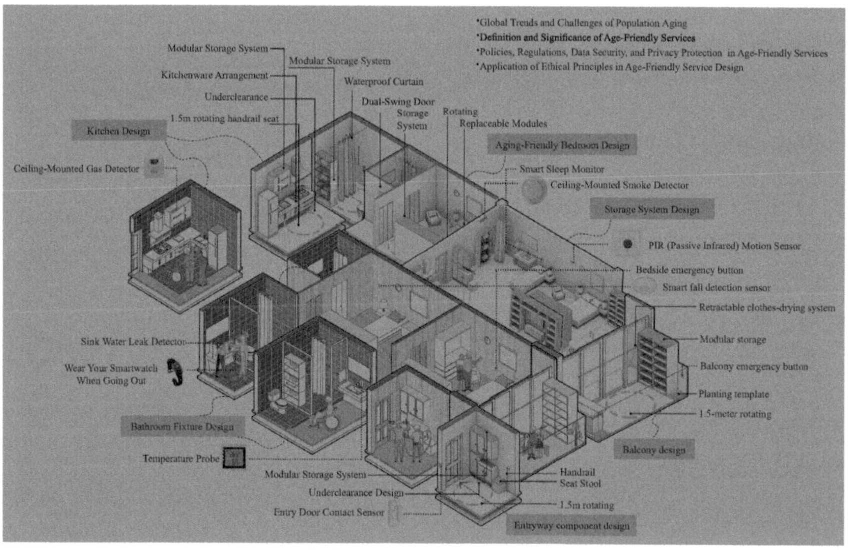


## 1 Global Trends and Challenges of Population Aging

### 1.1 The Global Aging Trend

#### 1.1.1 Current Status of Global Aging

James Hillman insightfully remarked, "The 21st century may not turn green with ecological consciousness, but it will undoubtedly turn gray with population aging." The global aging landscape in 2024 is particularly noteworthy, marked by a significant rise in the proportion of elderly populations across nations. While this trend is universal, its intensity and pace vary by country. According to the latest data from the United Nations and other authoritative sources, the population aged 65 and above is growing rapidly. In developed regions such as Japan and Europe, the elderly already constitute a persistently high and rising share of the population, profoundly reshaping socioeconomic structures. Meanwhile, populous nations like China, Germany, Japan, and the Russian Federation have either reached or are approaching population peaks, with elderly demographics claiming an increasingly prominent role.


W. Tang, *Aging-friendly Service System Design*, https://doi.org/10.1007/978-981-95-2054-1_1

### 1.1.2 Evolution of Global Aging

The global aging process is accelerating at an unprecedented rate. The United Nations Population Division projects that the population aged 65 and above will double within the next three decades, accounting for 16–22% of the global population by 2050. While regional disparities persist, aging remains an upward trajectory worldwide (see Fig. 1). Even in Africa, where demographic structures are relatively young, the growth rate of the 65–80 age group is twice that of Europe and three times that of North America. Additionally, between 2025 and 2054, 48 countries— including Brazil, the Islamic Republic of Iran, and Türkiye—are expected to reach historical population peaks. Although most nations that have already peaked are concentrated in Europe, Latin America and the Caribbean will emerge as the focal region over the next 30 years, with 19 countries accounting for 40% of total peaks. Looking further ahead, the global population is projected to peak at approximately 10.3 billion by the mid-2080s, a stark increase from 8.2 billion in 2024.

## 1.2 Challenges of Global Aging

### 1.2.1 Opportunities of Global Aging

The accelerating wave of global aging is profoundly transforming social, economic, and cultural landscapes, unlocking unprecedented opportunities in emerging sectors such as the "silver economy." As demographic structures shift, industries and services catering to older adults are expanding rapidly, with the silver economy becoming a pivotal driver of growth. Aging also stimulates the expansion of pension funds and diversification of commercial insurance products, providing long-term, stable capital inflows to financial markets and fostering multi-tiered development. This trend further accelerates the transition of industrial structures from labor-intensive to technology-, capital-,

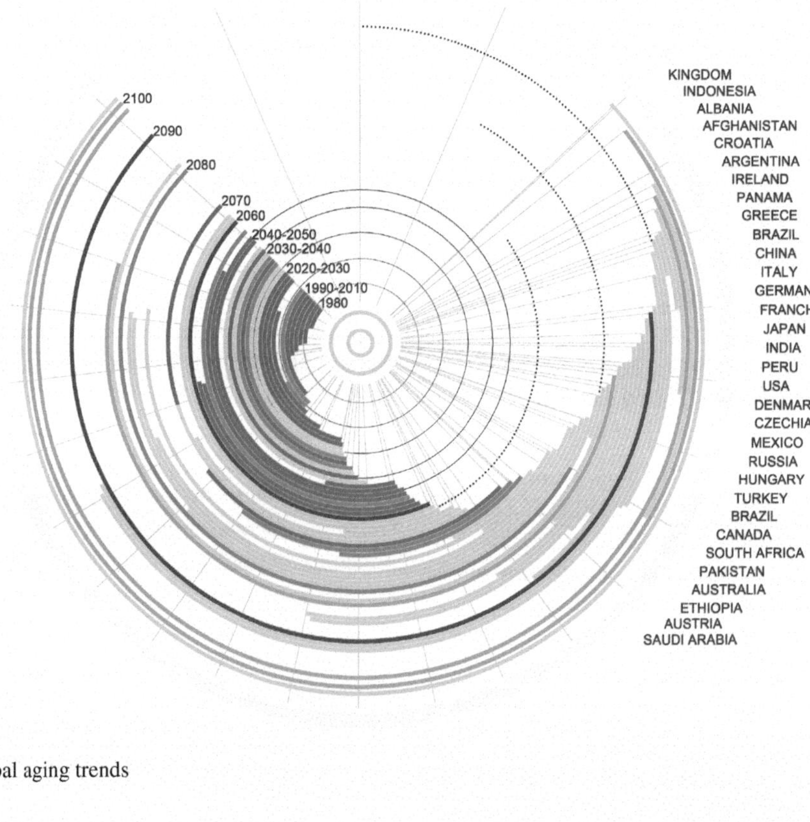


**Fig. 1** Global aging trends

and information-intensive models, injecting renewed dynamism into economic progress.

### 1.2.2 Crises of Global Aging

Global aging is exacerbating societal burdens through rising economic shares of non-productive populations, strained pension resources, and surging healthcare demands. This phenomenon significantly undermines potential economic growth rates and innovation capacity, posing challenges to labor supply, capital accumulation, and productivity—thereby constraining economic potential. The inevitable contraction of labor supply due to aging places acute downward pressure on growth prospects. Within social security systems, the escalating elderly dependency ratio intensifies pressures on public welfare and services, while the widening imbalance between pension supply and demand has emerged as a critical societal challenge demanding urgent resolution.

# 2 Definition and Significance of Age-Friendly Services

## 2.1 Age-Friendly Services

### 2.1.1 Conceptual Explanation of Age-Friendly Services

Age-friendly services are designed from the perspective of elderly users, tailoring service functions and formats to their needs. Services refer to interactive processes between users and organizations, through which benefits are derived. Often termed "intangible products," they encompass public attention, advice, methods, experiences, and emotional labor. Emerging against the backdrop of population aging, age-friendly services adopt a user-centered approach, applying service design as a systemic methodology to develop comprehensive solutions for elderly populations (see Fig. 2).

### 2.1.2 Research Evolution of Age-Friendly Services

In 1991, Bill Hollins and his wife formally introduced the term "service design" from a design perspective in their book Total Design: Managing the Design Process in Service Sectors. In 1994, the British Standards Institution (BSI) issued the world's first guidance standard for service design management, Design Management Systems: A Guide to Service Design Management. The UK's first service design company, Live|Work, was established in 2001. In 2008, Finland's Aalto University founded the Service Factory, and the Service Design Network (SDN) was officially launched, becoming a pivotal international platform for practitioners and scholars. In 2012, Professor Wang Guosheng of Tsinghua University organized the establishment of "SDN

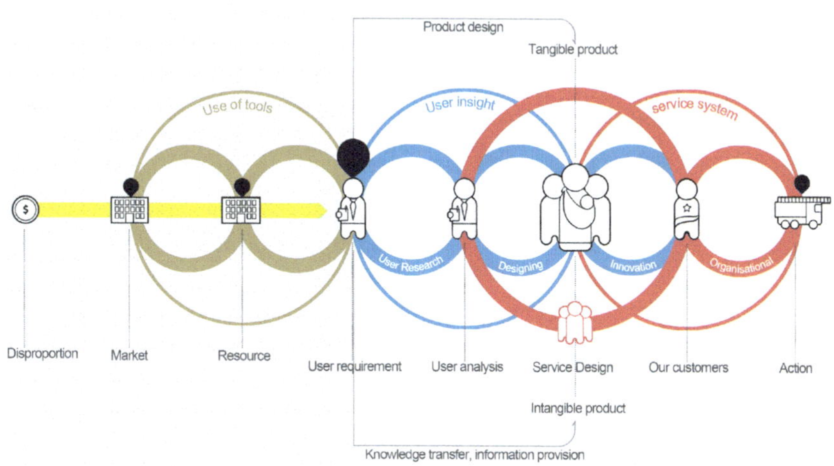


**Fig. 2** Definition of service design

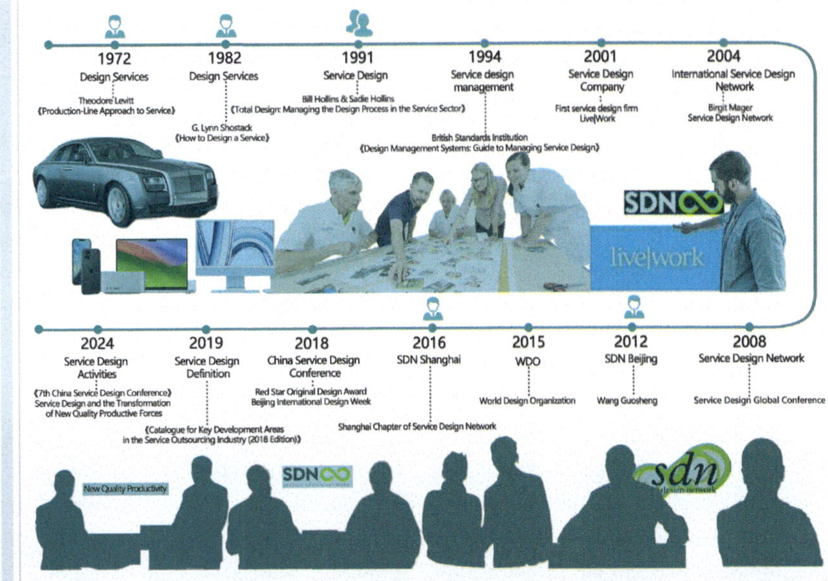


**Fig. 3** Evolution of service design

Beijing" to promote service design concepts in China. The China Service Design Development Research Center was founded in 2015. In 2016, Professor Wang collaborated with multiple institutions to launch the "Service Design Education and Research Forum," which continues to thrive. By 2024, the China Service Design Conference had successfully convened its seventh edition, solidifying its status as a premier event in the field (see Fig. 3).

## 2.2 Significance of Age-Friendly Services

### 2.2.1 Impact of Age-Friendly Services

Big Data-Driven Elderly Care Services

The advent of the digital-intelligent era has expanded the application of big data in age-friendly service design. By integrating data processing and algorithmic innovation, deep mining and analysis of data enable personalized, intelligent, and efficient service experiences for elderly users. Big data-driven design emphasizes contextual awareness and human-technology synergy, leveraging both human insight and technological support to enhance elderly care. For instance, the "Smart Walnut Health Ball" designed by Seohee Lee and colleagues adapts the traditional walnut hand-massage concept to monitor health metrics. By precisely capturing grip strength and hand movements, it collect s and analyzes health data comprehensively (see Fig. 4), offering robust support for elderly health management.

**Artificial Intelligence-Driven Elderly Care Services**

Powered by digital technologies, artificial intelligence (AI) has become a core driver for advancing age-friendly services, offering unparalleled advantages in content creation, contextual innovation, experience enhancement, and intelligent care. The rapid development of cloud computing provides AI algorithms with precise data support and robust computational power, ensuring smarter and more accurate services. AI-driven solutions not only keenly identify elderly needs but also dynamically adapt service methods, enabling efficient and targeted innovations in elderly care.

**Fig. 4** Smart Walnut Health Ball

#### 2.2.2 The Imperative of Age-Friendly Services

Current elderly care standards and systems remain underdeveloped, with age-friendly services facing practical challenges such as mismatches between products/services and elderly needs, as well as a shortage of specialized talent, which hinders service quality. Integrating age-friendly services as a systemic approach into public affairs and policymaking enables holistic solutions. By prioritizing user-centricity, this approach facilitates the construction of comprehensive elderly care ecosystems, ultimately elevating user experiences.

## 3 Policies, Regulations, Data Security, and Privacy Protection in Age-Friendly Services

### 3.1 Policies and Regulations for Age-Friendly Services

#### 3.1.1 Domestic Policies and Regulations

Amid accelerating population aging and the imperative of supply-side structural reforms, the Chinese government has prioritized addressing the challenges and opportunities arising from this demographic shift. Efforts to strengthen the management and legal framework for age-friendly services are intensifying, integrating proactive aging concepts and healthy aging principles into socioeconomic development. China is accelerating the establishment of a comprehensive policy system and institutional framework for age-friendly services, introducing more specialized talent and technological solutions to enhance efficiency and drive high-quality development in elderly care initiatives (see Table 1).

#### 3.1.2 International Policies and Regulations

As global aging intensifies, many developed nations have made significant strides in legislative and service system development, establishing relatively mature and efficient elderly care frameworks. These systems emphasize not only improving the quality of care institutions but also clarifying governmental and societal responsibilities through detailed laws and policies. This ensures comprehensive, high-quality, and compassionate services for older adults (see Table 1).

### 3.2 Data Security and Privacy Protection in Age-Friendly Services

#### 3.2.1 Data Security in Age-Friendly Services

The Data Security Law defines data security as "ensuring data remains effectively protected, lawfully utilized, and capable of sustaining a secure state through necessary measures." In the digital era, the rapid advancement of big data and cloud computing technologies has become a key driver of societal progress and industrial transformation. Given the profound implications of data security for daily life and production, it is imperative to prioritize data protection within big data ecosystems. This involves enacting relevant laws and regulations, constructing multi-layered, dynamic security architectures spanning physical, data, application, and network layers,

**Table 1** Domestic and international policies and regulations on age-friendly services development

| Type | | Name and content of policies and regulations |
|---|---|---|
| *China's policies and regulations on developing elderly-friendly services* | | |
| Notice | 2011 | **Notice of the General Office of the State Council on Issuing the Plan for the Construction of Social Elderly Care Service System (2011–2015)**<br>Proposed the requirement to "prioritize the development of social elderly care services", emphasizing the use of modern scientific and technological achievements to improve service management standards and quality |
| | 2017 | **Notice of the State Council on Issuing the 13th Five-Year Plan for National Aging Initiatives and Elderly Care System Development**<br>Aimed to enrich elderly care service formats, boost the senior product market, increase the supply of elderly-specific goods, enhance the technological sophistication of senior products, and strengthen R&D of age-friendly technologies |
| | 2022 | **Notice of the State Council on Issuing the 14th Five-Year Plan for National Aging Initiatives and Elderly Care Service System Development**<br>Advocated vigorous development of the silver economy, accelerated R&D, manufacturing, and intelligent upgrading of elderly care products, and promoted the application of smart health and elderly care |
| Opinions | 2016 | **Opinions of the General Office of the State Council on Fully Opening the Elderly Care Market to Improve Service Quality**<br>Called for full liberalization of the elderly care market, significant enhancement of home- and community-based elderly care quality, and comprehensive construction of a high-quality elderly care supply system |
| | 2019 | **Opinions of the General Office of the State Council on Promoting Elderly Care Development**<br>Focused on high-quality development of elderly care services, continuous advancement of the smart health and elderly care industry, and expanded application of information technologies in elderly care |
| | 2023 | **Opinions on Promoting the Construction of a Basic Elderly Care Service System**<br>Formulate and implement the basic elderly care service list, improve the supply capacity of basic elderly care services, and enhance the accessibility and convenience of basic elderly care services |
| *Policies and regulations on developing elderly-friendly services in foreign countries* | | |
| Law | 1963 | **Japan enacted the Welfare Law for the Elderly**<br>Introduced construction standards for elderly welfare institutions and facilities, established elderly care service facilities and a home-based nursing system |
| | 1965 | **United States enacted the Older Americans Act**<br>Strengthened elderly-friendly services, protected social welfare rights of seniors, and promoted healthcare support for the elderly |
| | 1980 | **Sweden enacted the Social Services Act**<br>The government built a comprehensive elderly care service system with diverse services and meticulous care to meet varied service demands |
| | 1994 | **Germany enacted the Long-Term Care Insurance Law**<br>Expanded community-based elderly care services, enhanced standardized management of care systems, and improved professional expertise in service delivery |
| | 2007 | **France implemented the elderly care plans Aging Well (2007–2009) and Senior Solidarity (2007–2012)**<br>Established the Silver Economy Association, incentivized the use of elderly care vouchers, supported the development of care institutions, and actively promoted related products and services |

and establishing robust key management systems to enforce encryption of sensitive data while regularly updating encryption strategies. Organizations must also implement stringent access control policies based on data sensitivity, fostering a secure, healthy, and sustainable big data environment.

### 3.2.2 Privacy Protection in Age-Friendly Services

Privacy protection entails safeguarding personal information from unauthorized access, use, disclosure, or misuse through rational policies and technical measures. With the proliferation of AI, social media, and smart devices, individual data collection has become unprecedentedly convenient—yet this ease also risks exposing personal privacy and sensitive data to inadvertent harvesting and storage in online databases. Privacy protection extends beyond preventing leaks to countering unauthorized predictions and exploitation of personal behaviors and preferences.

## 4 Application of Ethical Principles in Age-Friendly Services

## 4.1 Definition of Ethical Principles

### 4.1.1 Conceptual Explanation of Ethical Principles

Design ethics integrates ethical concepts with social relations, guiding design activities through human consciousness and appropriate methodologies. According to Ethics, the term "伦" (lún) refers to "categories" and "理" (lǐ) denotes "patterns," extending to structured and logical reasoning. Ethics, in this context, represents decision-making and behavioral standards based on a value system. Rooted in moral relationships, ethical principles transcend the boundaries of the material world (i.e., design activities), social relations, and intellectual domains to address contradictions between materiality and ethical ideals. These principles span micro- to macro-ethical levels, applying ethical theories and developmental laws to establish behavioral norms and social standards. Through intentional design, they seek to foster societal coexistence, equality, and moral education by creating artifacts that align with collective moral values (see Fig. 5).

### 4.1.2 Rational Reflection on Ethical Principles

Utility, economy, and aesthetics have long been regarded as the three foundational principles of design. However, historical evolution and practical scrutiny necessitate revisiting these principles, leading to the proposal of a fourth: ethical principles. Beyond theoretical analysis, ethical principles systematically organize intellectual traditions and delve into ethical practices, merging theoretical depth with pragmatic wisdom. Design ethics, as defined here, emphasizes morality-centered, human-centric approaches that use visual communication to cultivate orderly and harmonious lifestyles. Grounded in moral philosophy, ethical principles establish the intellectual rigor and moral imperatives of design. They advocate for humanized considerations in the design process, prioritizing virtue ethics (the realization of moral excellence) and normative ethics (ethical standards).

## 4.2 Application of Ethical Principles in Age-Friendly Services

### 4.2.1 Implementation of Ethical Principles in Age-Friendly Services

Ethical principles influence designers' value judgments and choices through self-regulation. By integrating moral awareness into design philosophies, practitioners consciously align their actions with ethical standards, thereby realizing the ethical value of design. Ethics operates as a societal "ought-to-be" framework, reinforced through education and evolving values to become a collective pursuit within social and

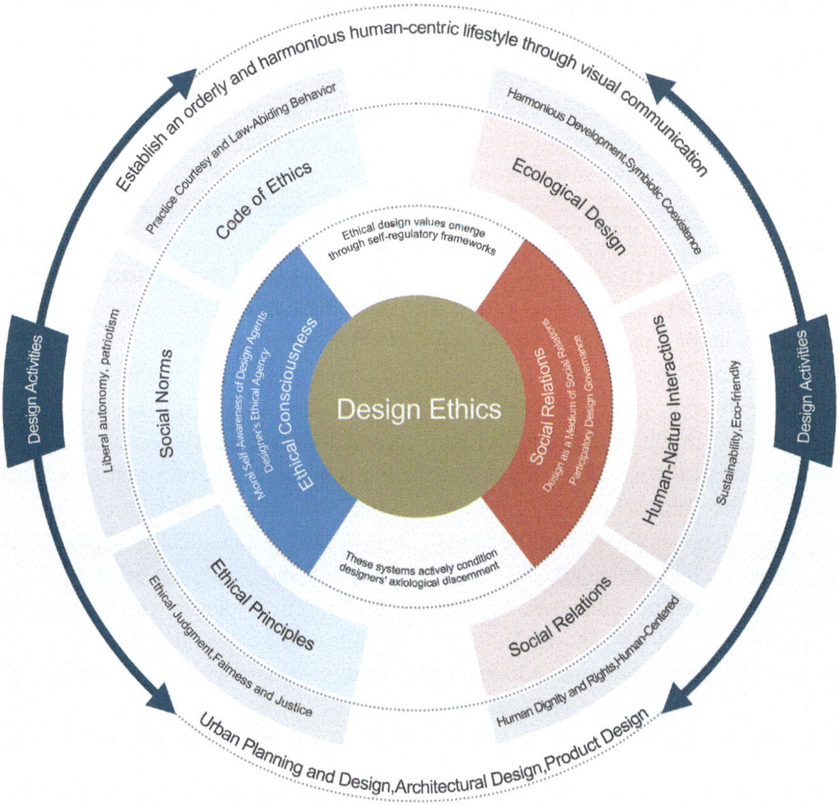


**Fig. 5** Definition of ethical principles

economic structures. To enhance ethical integrity, designers must elevate their moral literacy and professional ethics, ensuring compliance with ethically grounded standards.

### 4.2.2 Case Studies of Ethical Principles in Age-Friendly Services

Cobi, a detachable autonomous wheelchair designed by Hyeon Park, Hae un Jung, Hyun tae Kim, and Sookyoung Ahn, aims to foster positive lifestyle attitudes among older adults while challenging societal biases that often stigmatize them as burdens, particularly in the context of independent mobility. Equipped with autonomous navigation and an instant-braking onboard button, Cobi operates seamlessly without human intervention, enabling users to access areas unreachable by conventional transportation (see Fig. 6).

Addressing bathing challenges faced by older adults, especially those with limited mobility, Dongje Park developed the Cloud Shower System, a modular solution featuring three specialized components: a nylon brush for intensive scalp or foot callus cleaning, silicone bristles for massaging hard-to-reach areas like the back or scalp, and a soft sponge module for gentle cleansing integrated with an LED color-coded temperature control valve and one-touch dispensers for automated toiletry distribution (see Fig. 7). Both innovations exemplify user-centric design principles aimed at enhancing accessibility and dignity for aging populations.

**Fig. 6** Cobi detachable autonomous

**Fig. 7** Modular Cloud Shower System

# 2 Elderly Physio-Psychological Features Analysis

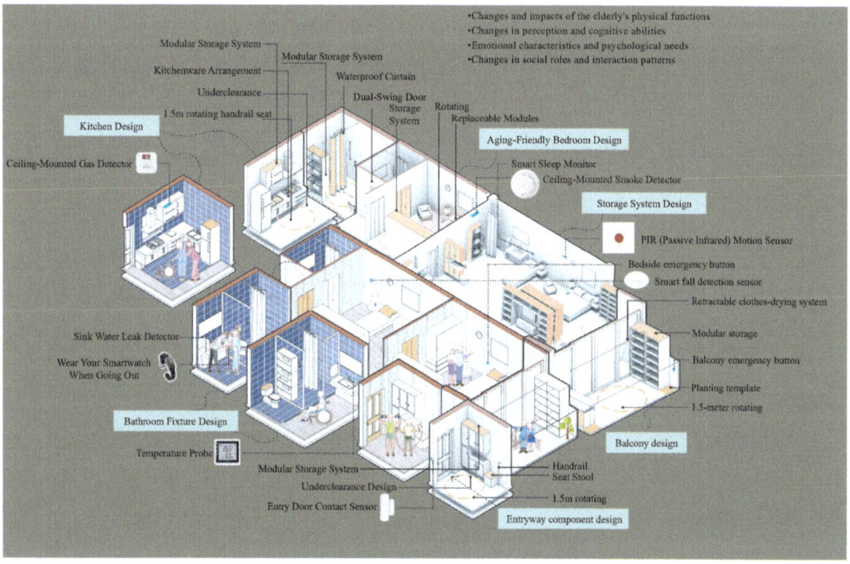


## 1 Changes and Impacts of the Elderly's Physical Functions

Naturally occurring ageing within the human body is a complex natural process that is continuous and progressive, not an accidental and sudden phenomenon. This process is accompanied by a gradual progression of physiological ageing in older persons, with changes in individual morphology and in the composition and functioning of the body's tissues and organs, which, to varying degrees, have far-reaching effects on the quality of life of older persons in their later years, leading to a gradual weakening of their physical and mental functions and an increasing risk of illness and death.

### 1.1 Physiological Changes and Impacts of Aging

#### 1.1.1 Characteristics of Changes in the Body Functions of the Elderly

Physiological aging, a natural phenomenon, is mainly manifested in the degenerative changes and functional decline of the eight systems of


W. Tang, *Aging-friendly Service System Design*, https://doi.org/10.1007/978-981-95-2054-1_2

the human body—the nervous system, immune system, circulatory system, respiratory system, digestive system, locomotor system, urinary system and endocrine system. This is a natural phenomenon in which a series of physiological changes in gene expression and cellular function gradually occur in terms of morphology, function and metabolism. These changes, though subtle and slow, accumulate into a huge impact that profoundly shapes the physical condition of the elderly.

### 1.1.2 Types of Physical Conditions in the Elderly

The diversity of the elderly state is determined by the significant differences in the physical functions and health levels of the elderly. In fact, the rate of deterioration of the elderly's physical functions is not completely proportional to age, but is affected by a combination of factors such as genetics, lifestyle and environmental conditions. It is a multi-dimensional consideration full of individual differences and uncertainties (see Fig. 1).

**Aging of the Body: Normal Process**

Physiological decline in the elderly is a natural, continuous, gradual process variably affecting all tissues and organs.

| System | Declining organs | Characteristics |
|---|---|---|
| Nervous | brain, spinal cord, cerebral nerves, spinal nerves, various ganglia | ↓brain weight ↓cells ↓blood flow and oxygen consumption ↓neurotransmitters mental retardation, forgetfulness, poor concentration, sleep issues, dementia |
| Immune | T-cells, B-cells | ↓new antigen-responsive leukocytes ↓complement proteins ↓lymphocyte conversion rate ↑susceptible to pneumonia, influenza, infective endocarditis, tumors ↑mortality rate |
| Circulatory | heart, blood vessels | myocardial fibrosis ↓excitability, conductivity, contractility; arrhythmia, dysfunction atherosclerosis, intimal thickness ↓elasticity ↓blood pressure and volume receptors |
| Respiratory | upper respiratory tract, trachea, bronchus, lungs | ↓lung ventilation/gas exchange ↓respiratory tract defense function ↑chest tightness, cough, obstruction, respiratory/pulmonary infections, respiratory failure |
| Digestive | oral cavity, esophagus, stomach, intestines, liver, gallbladder, pancreas, anus | ↓gingiva ↓saliva ↓chewing ↓appetite ↓gastrointestinal motility, iron malabsorption ↓liver size/enzyme activity ↓detoxification function ↓drug metabolism |
| Locomotor | bones, joints, muscles | ↑bone water ↓calcium carbonate, osteoporosis ↑back-leg pain, fractures, deformities ↑joint cartilage/ligament calcification-fibrosis ↓muscle/ligament atrophy reflexes, clumsiness |
| Urinary | kidney, ureter, bladder, urethra | renal arteriosclerosis ↓renal blood flow ↓acid, alkali and water electrolyte metabolism bladder muscle ↓capacity ↑urination, incontinence ↑drug toxicity sensitivity |
| Endocrine | hypothalamus, pituitary, thyroid, adrenal, thymus, gonads, pancreas | ↓endocrine gland stimulation response ↓pituitary weight ↓thyroid synthesis/secretion ↓stress response ↓gonadal activity ↓insulin secretion ↓functional sensitivity |

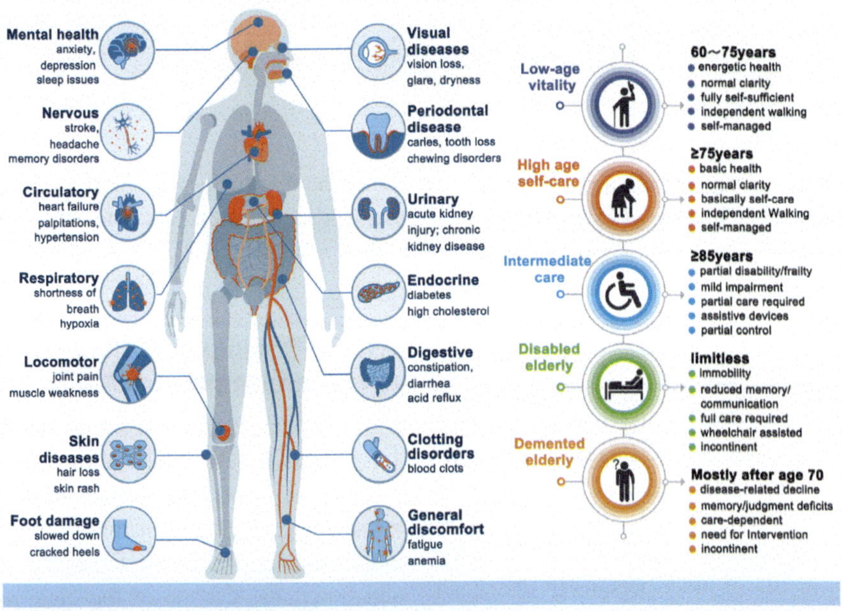


**Fig. 1** Characteristics and types of physiological changes in the elderly

1. Healthy elderly (60–75 years): Generally in good health, self-sufficient in daily life, and maintaining strong cognitive, communicative, and social engagement abilities. They lead active, fulfilling lives.
2. Independent elderly (≥75 years): Despite advanced age, they remain self-reliant, with stable chronic conditions. They retain clear cognition, mobility, and continence, demonstrating resilience and optimism.
3. Self-care elderly people with chronic diseases: This group of elderly people face many challenges. They may have problems with forgetfulness, weakened communication and judgment abilities, or physical disabilities, which lead to a certain degree of reliance on the assistance of others in their daily lives. They can use equipment to carry out outdoor activities. Although they need help to move, they can eat independently and control their bowel movements and urination, showing a tenacious spirit.
4. Bedridden elderly people: Due to long-term illness or disability, their daily living ability is seriously impaired, including long-term bed rest and wheelchair use. This type of elderly people partially or completely rely on the help of others. Their living environment is limited to indoors and they cannot go out to enjoy the sun and fresh air.
5. Elderly people with dementia: Dementia is a progressive cognitive decline caused by brain damage or disease. Its degree of degeneration far exceeds normal aging. The memory, attention, language and other abilities of elderly people with dementia are seriously affected. They may not be able to distinguish people, things, time and places, and fall into chaos.
6. Elderly people in professional care: This group of elderly people includes the elderly who are very old, seriously ill, postoperative, disabled, demented and terminally ill. They are at the most vulnerable stage of their lives and need to be provided with 24-h uninterrupted life care services. The professional care team will continuously adjust the service plan according to the changes in the elderly's physical condition to ensure that they are provided with professional care and health management services so that they can feel warmth and dignity in the last stage of their lives.

## 1.2 Pathological Changes of Aging

### 1.2.1 High Incidence of Chronic Diseases

Among adults aged 65+, 80% have at least one chronic condition, with many experiencing multiple comorbidities. Common diseases include hypertension, coronary heart disease, diabetes mellitus, chronic obstructive pulmonary disease, tumours and senile dementia. Due to the long duration of the disease and its protracted nature, they cause significant physical and mental harm to the elderly, affecting their ability to work and their quality of life (see Fig. 2).

### 1.2.2 Degenerative Organ Changes

When we enter the middle and old age, our body functions gradually decline, metabolism slows down or organ function decreases, which are all inevitable processes of human aging. Typical degenerative lesions include degenerative heart valve disease, senile cataract, degenerative osteoarthropathy, osteoporosis, Alzheimer's disease and Parkinson's disease. These lesions are associated with aging, long-term wear and tear, and decreased repair capacity. And for non-organic pathologies, maintenance and prevention are far better than over-treatment.

### 1.2.3 Acute Illness Susceptibility

When an elderly person suffers from a certain disease, he or she is often prone to complications on the basis of the original disease, a phenomenon closely related to the coexistence of multiple diseases, the decline in immune function, and the weakening of the ability to

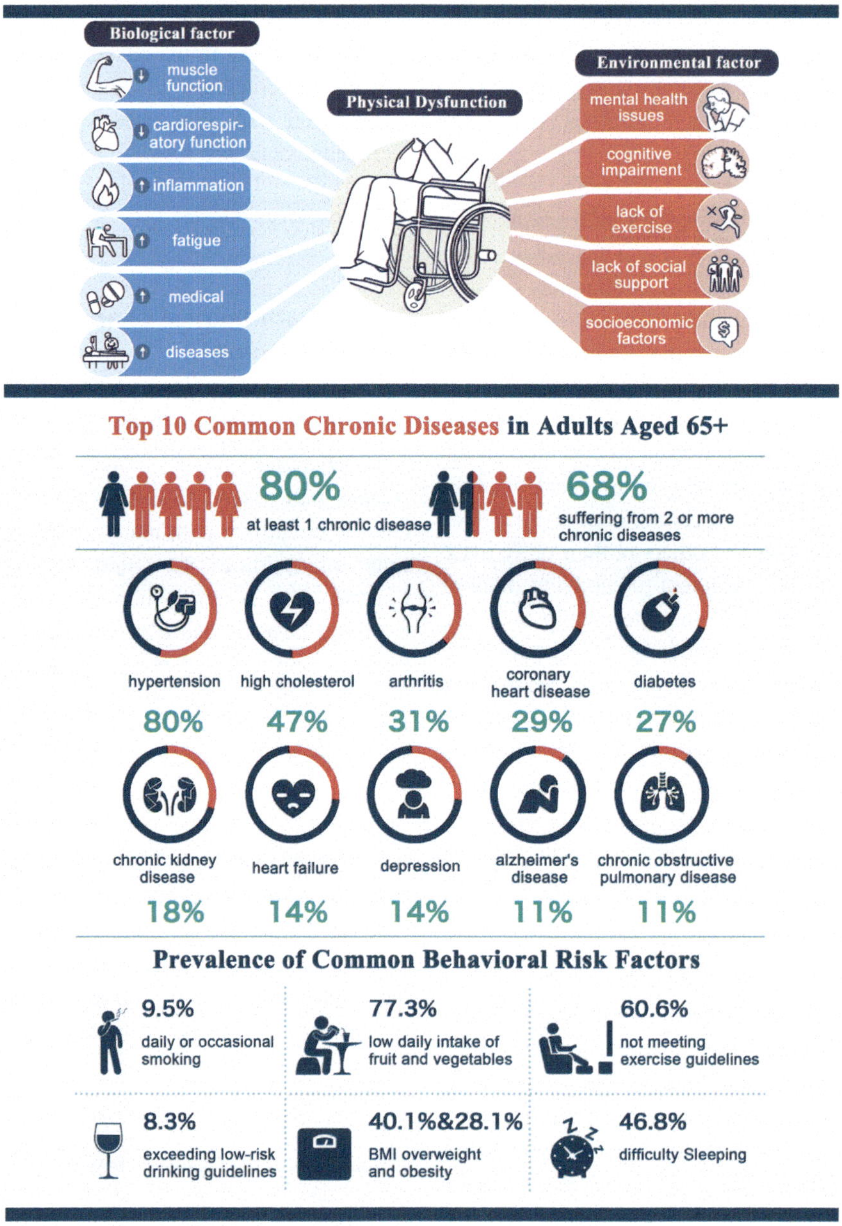


**Fig. 2** Chronic diseases and risk factors in the elderly

resist stress reactions in the elderly. Common complications of acute diseases include acute pneumonia, urinary tract infection, myocardial infarction, malignant tumour and uremia. The emergence of such diseases is usually sudden and should be detected and treated early (see Fig. 3).

## 2 Changes in Perception and Cognitive Abilities

Perception is the process of the brain's reflection of objective things in the real world, while cognition is the process by which individuals acquire knowledge and information through

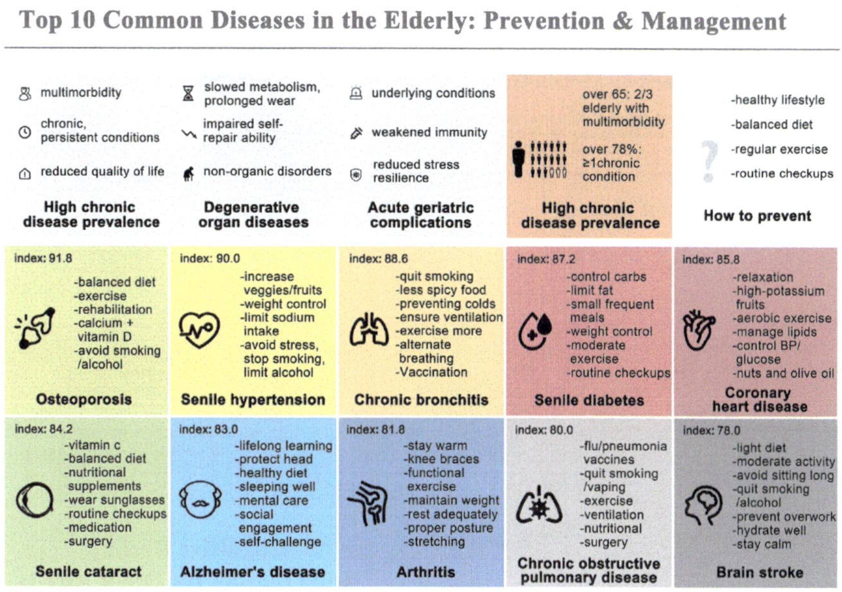


**Fig. 3** Top 10 common geriatric diseases and prevention

concept formation, perceptual understanding, judgemental reasoning and imaginative creation. Age-related declines in these abilities often trigger negative emotions, creating social, psychological, and daily living challenges (see Fig. 4).

## 2.1 Perceptual Changes in the Elderly

### 2.1.1 Vision

Reduced corneal diameter, reduced perceptual sensitivity, and ageing of the vitreous and retina and other eye structures in the elderly lead to a decline in bright and dark vision, sensitivity to glare, weakened colour discrimination, and slower processing of visual information by the brain. In addition, the probability of the elderly suffering from pathological changes such as cataract, macular degeneration, glaucoma and diabetic retina is also greatly increased.

### 2.1.2 Hearing

The hearing system of the elderly, from the endings of the cochlea to the auditory centre shows a trend of gradual decline, its auditory function deteriorates, the clarity of auditory perception decreases, and the ability to perceive sound decreases. Pathological aspects leading to hearing loss in the elderly are mainly caused by senile deafness, otitis media and drug-induced deafness, etc., which have a certain impact on the interpersonal communication, physiological health and quality of life of the elderly.

### 2.1.3 Taste

Taste perception relies mainly on the tongue, and the reduced number and sensitivity of taste buds in the elderly is an important factor in the decline of their taste function. Reduced salivary secretion and reduced chewing ability in the elderly have different effects on their appetite and taste. At the same time, the type of food, temperature, personal health conditions and poor living habits can also cause the sense of taste of the elderly to become dulled and their ability to distinguish between various flavours to decline.

### 2.1.4 Smell

Ageing is a major cause of olfactory function decline in the elderly. Long-term working and living environments, air circulation, personal oral hygiene, dietary habits and diseases caused by physiological factors all have an impact on the olfactory function of the elderly.

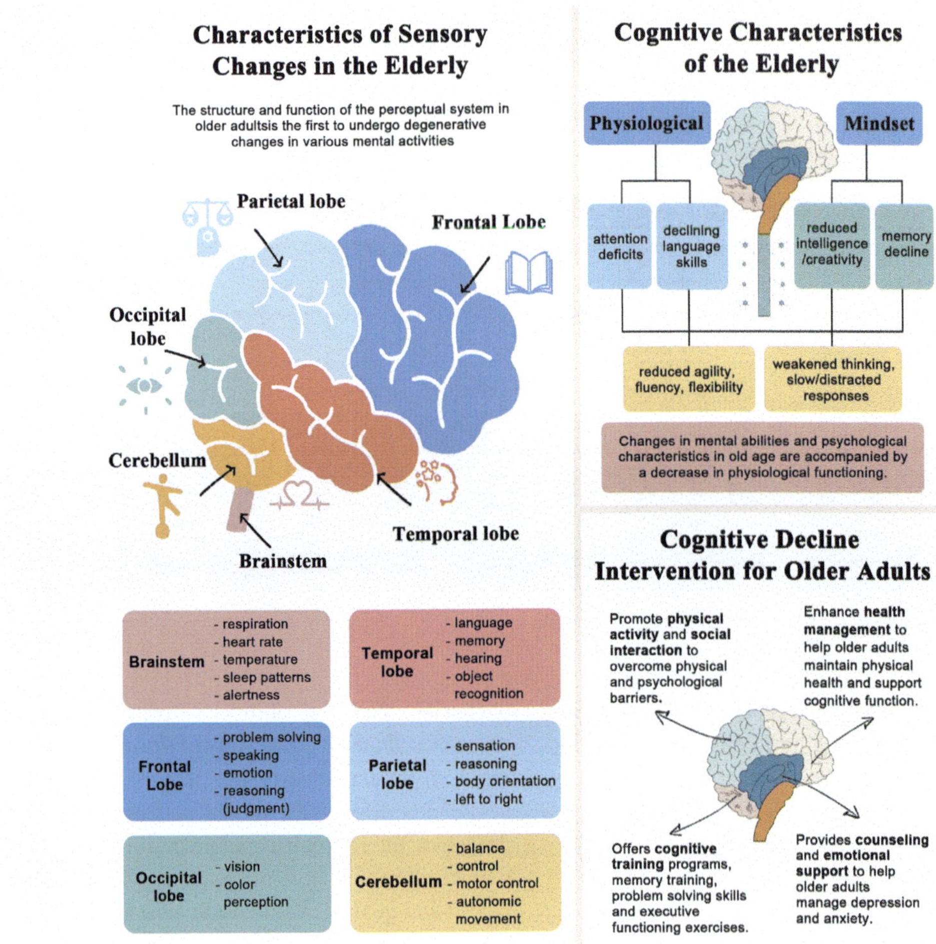


**Fig. 4** Characteristics & impacts of reduced perceptual and cognitive abilities in seniors

### 2.1.5 Touch

Skin sensations mainly include touch, temperature and pain. The number of tactile points on the skin of the elderly gradually decreases with age, resulting in a continuous increase in the skin's need for tactile stimulation intensity. The decline in skin sensory function can easily lead to skin injuries such as bumps and burns.

## 2.2 Cognitive Characteristics of Aging

### 2.2.1 Attention Decline

Reduced attention and concentration in older adults is associated with structural changes in the brain, leading to difficulties in concentrating and staying mentally engaged, which manifests itself in forgetfulness and cognitive decline, with symptoms varying from person to person. Therefore, the elderly will feel frustrated and experience negative emotions such as depression and anxiety.

### 2.2.2 Intelligence & Creativity

Intelligence is a comprehensive cognitive ability that includes observation, attention, memory, imagination and thinking. Changes in intelligence in the elderly do not manifest themselves as a decline in all factors together, but rather as an imbalance in the changes in the factors, a decline in the ability to think out of the box, and a faster and more significant decline in motor intelligence than in verbal intelligence in the elderly.

### 2.2.3 Memory Impairment

Memory loss in the elderly is more pronounced and sensitive, which can impair their quality of life and work efficiency and increase the risk of Alzheimer's disease. The health status, gender, psychological quality, occupational attributes, and hobbies of the elderly all affect their memory, so it is crucial to take scientific and effective preventive measures in their daily lives.

### 2.2.4 Language Challenges

The decline in the oral expression ability of older persons, such as forgetting words, reduced syntactic complexity, lack of focus in conversation s or a greater tendency to stray from the topic, and a decline in the ability to read and understand or express themselves in writing, is a result of physiological ageing. Language, as an important carrier for individual development, interpersonal communication and social participation, requires the concerted efforts of the individual, the family and the society to effectively solve the language problems of the elderly.

## 3 Emotional Characteristics and Psychological Needs

Due to differences in life experiences, cultural backgrounds, living environments, personality traits, and behavioral needs, the elderly enter the senior stage in diverse and unique ways, leading to varying emotional characteristics and psychological needs (see Fig. 5).

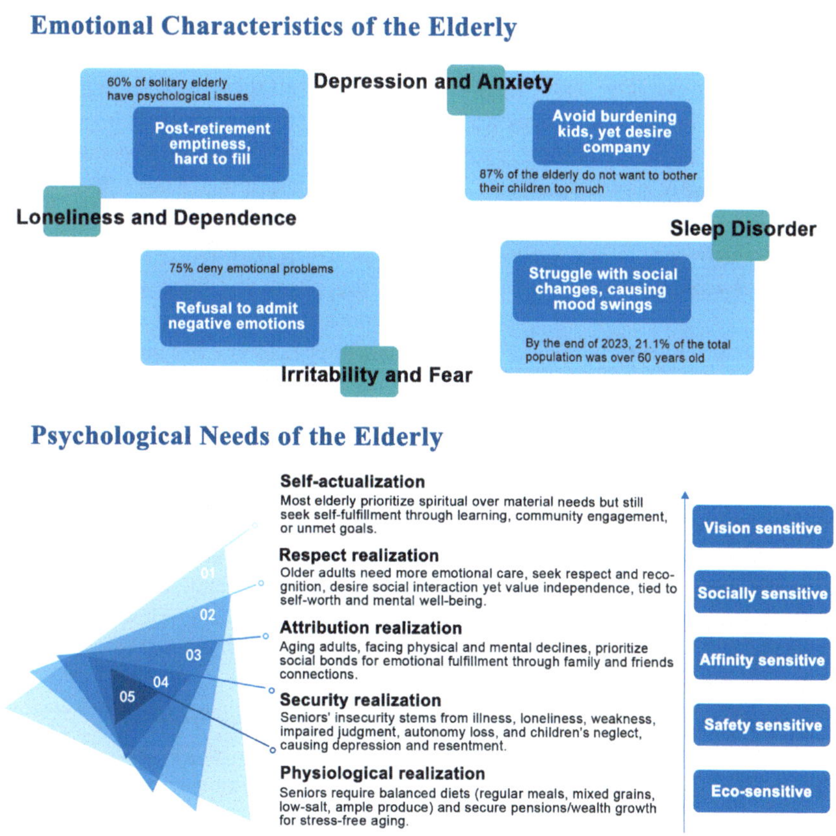


**Fig. 5** Analysis of elderly emotional traits and psychological needs

## 3.1 Emotional Traits of the Elderly

### 3.1.1 Loneliness & Dependence

As they grow older, older people may experience widowhood, living alone, retirement, reduced mobility, fewer contacts, and changes in their social and family status, etc. These changes lead to a gradual loss of a sense of control and initiative over their lives, and they can only passively adapt to and submit to the arrangements of others. This leads to feelings of helplessness, loss, isolation, and loneliness. Meanwhile, they lack confidence, are emotionally fragile and indecisive. Excessive mental dependence causes emotional swings and sensory decline, making them eager for understanding and support.

### 3.1.2 Irritability & Fear

When the elderly experience change s in their endocrine and neurotransmitter systems, as well as challenges to their social status, economic situation and family relationships, they often feel that they are unable to adapt to the rapid development and changes in society. This sense of inability to adapt, coupled with concerns about illness and declining self-care ability, causes them to be psychologically worried and fearful, and to display negative emotions such as apathy and impatience, which can cause considerable distress in their later years.

### 3.1.3 Depression & Anxiety

Some older persons are plagued by chronic illnesses and face the threat of death, which makes them prone to fear and depression. Some older persons suffer from anxiety and depression as a result of monotonous life, widowhood, escalating family conflicts and spiritual emptiness. Some older persons are depressed and discouraged by the drastic changes in their lifestyles after retirement, the reduction in social interaction and the lack of a sense of belonging and identity, resulting in depression, discouragement and a loss of confidence and hope in the future.

### 3.1.4 Sleep Disorders

The elderly commonly have trouble sleeping, with some having difficulty falling back asleep after waking, experiencing symptoms like insomnia and vivid dreams. These sleep disorders are not only related to the decline in the excitatory and inhibitory abilities of the cerebral cortex in the elderly, but may also be a complication of diseases such as brain lesions. Chronic sleep deprivation can seriously affect the physical and mental health and quality of life of the elderly.

## 3.2 Psychological Needs of the Elderly

### 3.2.1 Dependence Needs

When the elderly retire, the family becomes their main place of activity and spiritual support. However, old age and frailty and the reality of their children starting their own families make them feel inactive in their lives and psychologically long for more care and attention. Especially in their dealings with outsiders, they rely on and look forward to such support and care, and this need for dependence is an indispensable part of their later life.

### 3.2.2 Help Needs

As the physical fitness and self-care As older people's physical fitness and ability to care for themselves deteriorate, they need more and more help and care from others. Whether it is simple daily living or more complicated health management, they are increasingly unable to cope. If these needs are not met in a timely and effective manner, the elderly may develop negative emotions such as depression and resentment, and may even feel abandoned. This need for help is like sweet rain in the midst of a drought, and is the psychological comfort they desperately need.

### 3.2.3 Self-Esteem Needs

Older persons who are retired or incapacitated change their social role from provider to dependent. Although they realise that they are older and not as capable of working or earning as much financially as they used to, they still want to be seen as valuable individuals, respected and understood by others. This need for self-esteem allows them to maintain dignity and self-confidence in their later years. They do not want to be treated as an invalid, but want to continue to fulfil their role and value in society and in the family.

## 4 Changes in Social Roles and Interaction Patterns

### 4.1 Elderly Role Changes

Old age is life's final significant transition. Retirement, its most prominent aspect, alters long—established activities and social roles, thus changing elders' life content and behavior (see Fig. 6).

#### 4.1.1 Shift from Professional to Leisure Roles

After retirement, older persons are gradually separated from their original occupations and social life, and the change of roles has a greater impact on their lives and psychology. Work is not only the main source of income for older persons, but also an important way for them to gain a sense of satisfaction, fulfilment and achievement. Retirement means breaking the specific lifestyle and habits of older persons, which can make them feel confused and at a loss as to what to do.

#### 4.1.2 Transition from Primary to Dependent Role

Before their retirement, the elderly had their own jobs, relationships and stable financial incomes, and were the dominant players in their families, not to mention an important platform

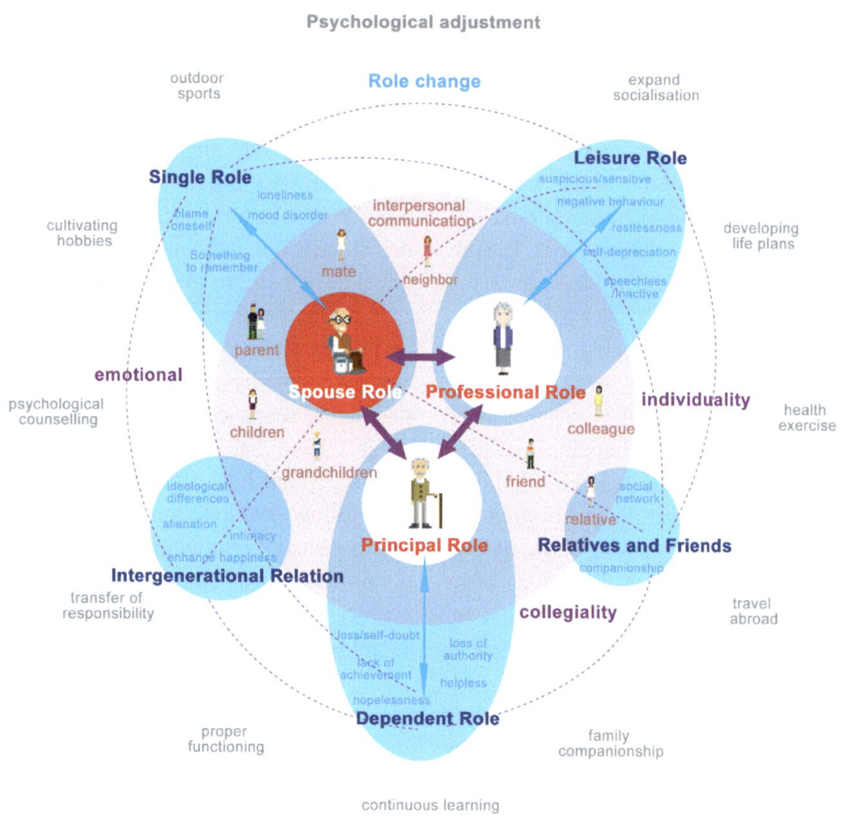


**Fig. 6** Elderly interaction pattern changes and impacts post—role transition

for them to realise their self-worth and maintain their social ties. After retirement, they have shifted from being relied on by their children to being dependent on them, which to a certain extent affects their self-esteem and self-confidence, and they lose their original dominant role and sense of authority in the family.

### 4.1.3 Change from Spousal to Single Role

In old age, the likelihood of widowhood increases, with couples rarely dying on the same day, and once a spouse has died, the other spouse has to suddenly enter the role of single life. Adjusting to this change is often a challenging adaptation process, and older persons may experience deep anxiety due to fear of facing their future lives alone, concerns about their health, or missing their deceased spouse.

## 4.2 Components of Elderly Interpersonal Relationships

First, family relationships. The spousal bond lies at the heart of the family, enabling the fulfillment of numerous family functions. Most elderly couples enjoy stable, harmonious, sincere, and deep—seated relationships. Despite occasional bouts of melancholy due to empty—nest syndrome or squabbles over trifles, they generally communicate amicably to resolve issues.

Second, intergenerational relationships. When the elderly evaluate the happiness of their later life, whether their children are filial is a crucial indicator. Whenever their children are filial, the elderly will have a strong sense of happiness; if their children are not filial, even if the elderly have relatively abundant material conditions, their sense of happiness will be greatly reduced.

Third, relationships with relatives and friends. As physical capabilities wane, including reduced mobility and slower reactions, elders, driven by shared interests, complementary needs, and similar attitudes, tend to engage with like—minded individuals through activities such as dancing, chess—playing, and fishing.

## 4.3 Characteristics of Elderly Interpersonal Communication

First, the elderly show distinct individuality in interpersonal communication. After retirement, the elderly are more free and easy, and the individual characteristics in interpersonal relationships become particularly prominent. The main place of activity for the elderly has shifted from work to society, community and family. They need to play other social roles to establish their social position and find a sense of belonging and meaning in the new social environment.

Second, the elderly show significant group nature in interpersonal communication. The elderly should strengthen their understanding of group nature and actively participate in cultural activities, social gatherings, etc. Not only can they cultivate interests and make their mood happy, but they can also gain a sense of belonging. Providing new social contact channels for the elderly not only meets the actual needs of the elderly, but also meets their spiritual needs.

Finally, the elderly are also rich in emotions in interpersonal communication. Because most elderly people have mature and sound personalities, noble sentiments, are polite in dealing with others, have self-esteem and self-confidence, cherish friendship, are constrained by society, and follow group norms. Therefore, through interpersonal communication, the elderly can obtain emotional satisfaction and support, which is very important for their mental health and social adaptation.

# 3 Design Principles of the Aging-Friendly Service System

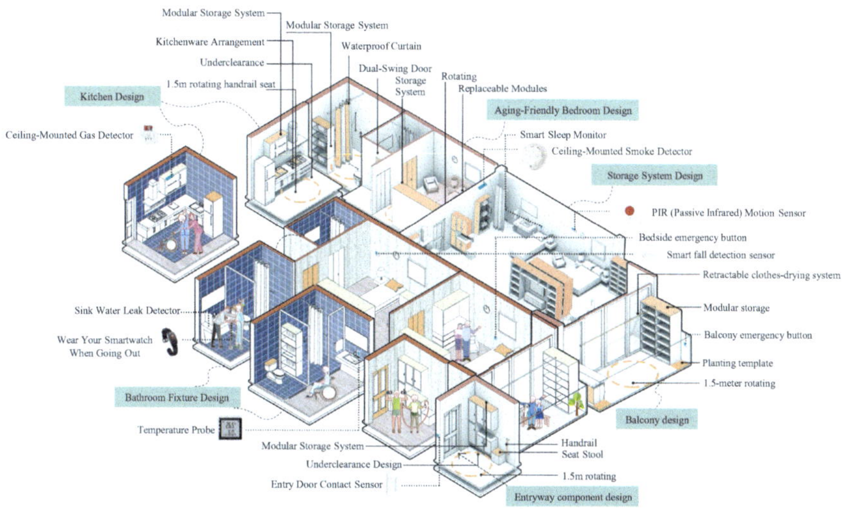


## 1 Principles of Universal Design

Universal design principles aim to create products and environments that are accessible to everyone. It emphasizes accessibility, flexibility, simplicity intuitiveness, and consideration of diverse needs to ensure broad usability and inclusivity.

### 1.1 Definition and Significance of Universal Design

#### 1.1.1 Definition of Universal Design

Professor Pulos of the United States once said, "The purpose of design is neither profit making nor inflating the glory of designers, but rather to


W. Tang, *Aging-friendly Service System Design*, https://doi.org/10.1007/978-981-95-2054-1_3

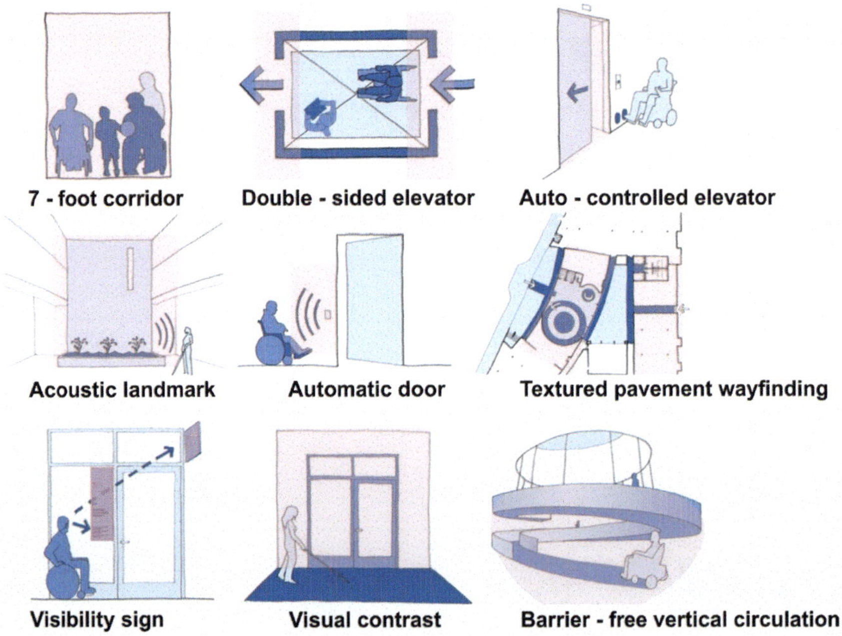


**Fig. 1** The embodiment of universal design

serve humanity with civilized design as much as possible." This view reveals the core mission of design, that is, to meet the service related needs of people. Universal design is precisely based on this concept. In the design of products, environments, buildings, and services, it takes into account the diversity of all people throughout the process to ensure that the design outcomes can be used fairly and effectively by everyone, regardless of age, ability, or background.

Universal design not only focuses on accessibility but also emphasizes the flexibility of products and environments, simple and intuitive ways of use, and the ability to adapt to a wide range of user groups. Its goal is to eliminate barriers in design so that both the elderly, the disabled, and the general public can use these design results independently, safely, and comfortably without special adaptations (see Fig. 1).

### 1.1.2 Significance of Universal Design

#### Human-Centered Spirit

During the product development and design process, designers need to consider various user groups and incorporate their lifestyles into the scope of consideration and attach importance to them. Universal design emphasizes putting human needs at the core, focusing on the perspective of users to think, so that designed products can meet the needs of more diverse life patterns and create higher value. Its essence and core concept are "human-centered design in all aspects" (see Fig. 2).

#### Psychological Care

"Care" is the key to universal design. By meeting people's psychological needs and giving people a sense of respect, through creating accessible and user-friendly products and environments, universal design conveys care and respect for all users, especially for those groups often neglected in traditional design, such as the elderly and the disabled (see Fig. 2).

#### Easy to Use and Adaptable

The principles of easy-to-use and adaptability are inherent in things. Regardless of age, ability, or background, simple and intuitive designs can be understood and used. Universal design improves the usability of products and environments, reduces user thresholds, and makes it

**Fig. 2** The significance of universal design

easier for more people to safely use these design results, thus maximizing the user group (see Fig. 2).

#### Promote Social Inclusivity

By creating products and environments friendly to all people, it helps to eliminate social inequalities, especially discrimination against groups such as the disabled and the elderly. It encourages society to consider everyone in a more inclusive way, thereby promoting social fairness and overall-harmony (see Fig. 2).

#### Enhance Product and Environmental Sustainability

Considering the needs of a large number of users in the initial stage reduces the need for later modifications and adaptations. This not only saves—resources and costs but also extends the service life of products and environments, enhancing their sustainability. Universal design not only meets current user needs but also creates greater value for the future, reflecting a deep understanding of environmental and social responsibilities (see Fig. 2).

#### Enhance User Experience and Satisfaction

Through providing simple, intuitive, and easy-to-use products and environments, the overall user experience is improved and user satisfaction is increased. At the same time, it expands the user group, enhancing the competitiveness of products in the market. It can help the elderly use communication and social devices more easily, and also help the disabled move more freely in buildings, enabling all groups in the public to participate in social activities, so that everyone can feel equal treatment and greatly improving satisfaction (see Fig. 2).

## 1.2 Key Points of the Principles of Universal Design

The principles of universal design have been refined over time and are now quite mature. The most cited are the seven principles from the Center for Universal Design at North Carolina State University's research institute.

### 1.2.1 Equitable Use

The core is equal access. Products and services should be usable by all, regardless of ability, age, or background, without extra adaptation. This caters to diverse users, For instance, OXO GOODGRIPS creates kitchen utensils that can be easily used by people with different hand sizes and conditions, whether they are left- or right-handed.

### 1.2.2 Flexible Use

Designing for flexible use requires considering the diverse preferences, habits, and capabilities of users. It should allow for personalized usage, be convenient for both left- and right-handed individuals, and provide multiple operation modes. An example is IDC's medical glue pen. Its design, resembling a familiar pen shape, makes it easy for users to hold and operate.

### 1.2.3 Simple and Intuitive

Simple and intuitive design enables users to quickly understand and use a product. By ensuring that operation processes are clear and straightforward, it reduces the cognitive burden on users. "Double-sided Scissors" is a prime example. They can be used by both left- and right-handed people, and switching the blade direction is a simple operation.

### 1.2.4 Perceptible Information

Products should convey information through multiple sensory channels. Take the packaging of "M&M's" as an example. Its vivid colors and the easy-to-tear feature can be easily perceived by people with different visual and physical abilities, including children and the elderly with limited hand dexterity.

### 1.2.5 Tolerance for Error

Anticipating and preparing for user errors is crucial. By incorporating safety features such as undo options and safety locks, as seen in interface buttons and charging systems, the potential risk s and consequences of mistakes can be minimized, thus enhancing user safety.

### 1.2.6 Low Physical Effort

The goal is to reduce the physical and mental energy users need to expend. Products like the Dyson hand dryer are designed to operate automatically and efficiently, minimizing the effort required from users, which is especially beneficial for the elderly and those with limited physical strength.

### 1.2.7 Appropriate Space and Dimensions

Design s must accommodate people of all body types and mobility levels. Adequate space should be provided to ensure that users, whether they are in wheelchairs or standing, can use products or services comfortably and interact with them effectively. IDC incorporates scale considerations into medical device design based on in-depth research (Fig. 3).

## 1.3 Convergence Points of Universal Design and Age-Friendly Services

### 1.3.1 Universal Design and Age-Friendly Geriatric Care Products

Universal design in geriatric care products aims to boost the elderly's independence and comfort. It considers their physical and mental traits, creating easy-to-use, functional products.

Examples include adjustable hospital beds and automated aids. These have simple controls and suit various users. Such designs let the elderly be more self-reliant during rehab, cutting caregiver dependence and enhancing life quality.

### 1.3.2 Universal Design and Age-Friendly Visual Interfaces

Guided by universal design, visual interfaces for electronic devices help the elderly. They enlarge fonts, boost contrast, and simplify steps for their declining vision and cognition.

Voice and tactile feedback are added for more support. This eases device use, promoting the elderly's digital engagement and social integration.

### 1.3.3 Universal Design and Age-Friendly Spatial Environments

For age-friendly furniture, universal design targets ergonomic, versatile pieces. Electric lift-recliner chairs with easy-to-use remotes help the elderly stand and sit.

**Fig. 3** Seven principles of universal design

Adjustable-height tables meet different activity needs. Natural-material sofas offer comfort and safety. These improve the elderly's living quality.

### 1.3.4 Case Analysis: Successful Practices of Universal Design

The goal of universal design in age-friendly furniture design is to provide furniture products that are ergonomic and widely applicable for the elderly. With good operability and flexibility in material selection, first of all, the electric lift-recliner chairs adopt a simple remote-control operating system, allowing the elderly to stand up or sit down easily, improving the independence of daily life.

Secondly, the adjustable-height tables allow users to adjust the tabletop height according to the needs of different activities, meeting the diverse requirements of the elderly during dining, writing, or using a computer. Finally, the solid wood sofas (SolidWoodSofas) made of natural materials not only use environmentally-friendly wood, but also have excellent support and comfort, creating a healthy and safe home environment and avoiding potential threats to the health of the elderly from harmful chemicals. Evidently, the combination of universal design and age-friendliness significantly improves the quality of life of the elderly.

## 1.4 Case Analysis: Successful Practices of Universal Design

### 1.4.1 Practice 1—OXO GOODGRIPS

A series of kitchen utensils developed by OXO Company. The design concept of OXO

**Fig. 4** OXO good grips case study

GoodGrips is based on universal design principles, aiming to provide comfortable and easy-to-use kitchen utensils for all users, especially those with hand grip problems. Its most prominent feature is the emphasis on simplicity and intuitiveness, reducing potential user errors during use.

The knife tools have ergonomic soft-grip handles, ensuring that users can hold and use them easily in all conditions. In addition, the products cover all aspects of the kitchen, including bottle openers, peelers, measuring cups, etc. (Figs. 4 and 5).

## 2 Basics of Accessible Design

### 2.1 Definition and Development of Accessible Design

#### 2.1.1 Definition of Accessible Design

Accessible Design refers to considering the diverse needs of all people during the design of products, environments, buildings, and services. It aims to ensure that everyone, regardless of their physical ability, age, gender, or background, can use them fairly and without barriers. This design concept stems from society's concern for special groups such as the disabled and the elderly, with the goal of eliminating traditional design barriers and enabling these groups to participate in social activities independently and with dignity.

**Fig. 5** OXO good grips case study

This design covers physical environments (such as indoor and outdoor passages, restroom facilities, etc.), industrial products, and communication technologies (such as subtitles, screen readers, etc.). It also involves making services barrier-free.

### 2.1.2 Development of Accessible Design

Accessible design began with twentieth-century humanitarian ideas. In 1961, the US set the first "accessible standard", a global foundation. The Americans with Disabilities Act (ADA) later defined barrier-free needs in public buildings, transport, etc.

China has advanced barrier-free construction. It issued the Code for Accessible Design in 2012, Public Information Graphical Symbols: Symbols for Barrier—Free Facilities in 2017, and the Construction Law of the People's Republic of China on Barrier—Free Environment in 2023.

Accessible design then spread to industrial products and electronics. In the 1970s, the US and Europe set product-accessibility standards for ease-of-use. Japan led in accessible electronics, simplifying interfaces and adding aids.

In the twenty-first century, electronic info accessibility became key. Countries made relevant rules. The US required screen readers on government sites; Europe used EU ICT standards; China issued Guiding Opinions on Promoting Information Accessibility in 2020. Overall, it offers special groups more equal living and working chances.

## 2.2 Standards and Requirements of Accessible Design

### 2.2.1 Design for Specific Individuals

When designing for specific individuals, the key lie s in the in-depth fit with the usage situation. It is extremely important for users with disabilities to participate in the design process and express their personal needs. Designers should establish an intuitive understanding through their daily lives, and the designed solutions should precisely match current and future possible needs, taking into account factors such as people's aging or changes in mobility. Attention should be paid to the design's flexibility to ensure it can adapt to individual situations in the long term (see Fig. 6).

### 2.2.2 Design for Specific Groups

When designing for special groups, such as buildings for kindergartens, elderly care institutions, and special education schools, it is necessary to not only focus on the general needs of the target group but also consider non-dominant users. When dealing with multiple disabilities, especially when the needs of disabled people become highly individualized, even if adjustments are made for the main types of disabilities, it is difficult to create a completely barrier-free environment. Designers need to balance various needs.

### 2.2.3 Design for Non-specific Users

For places with non-specific users, such as outdoor public spaces, transportation spaces, and various public buildings, the design principle is to be barrier-free for as many people as possible. These places should accommodate people with disabilities who have a certain degree of independent mobility. However, due to the diversity of users, removing barriers for some may create new ones for others.

This requires designers to comprehensively consider various factors and weigh the pros and cons during the design process to maximize the overall barrier-free effect. In this way, the places can serve different types of users to the greatest extent, promoting equality and inclusivity in public spaces.

## 2.3 The Convergence of Accessible Design and Age-Friendly Services

### 2.3.1 Environmental Accessibility

By eliminating physical barriers, such as installing accessible handrails, accessible kitchens, and accessible elevators, etc., it ensures that the elderly can move independently and safely in public and private spaces (see Fig. 7) factors and weigh the pros and cons during the design process to maximize the overall barrier-free effect. In this way, the places can serve different types of users to the greatest extent, promoting equality and inclusivity in public spaces.

### 2.3.2 Transportation Accessibility

Upgrade public facilities such as low-barrier ramps, floor-level buses, and barrier-free subways.

Provide audible and visual—alert ticket machines. Ensure ramp slopes and widths meet standards and have anti-skid surfaces for easy self-use (see Fig. 7).

### 2.3.3 Digital Product Accessibility

Create simple, intuitive interfaces with larger fonts and icons, and voice prompts. Equip digital devices with assistive tech for elderly with special needs (see Fig. 7).

### 2.3.4 Information Accessibility

Guarantee the elderly can access and understand info. Use clear language, proper font sizes, and multi-sensory aids. Consider elderly-specific needs in public info dissemination (see Fig. 7).

## 2.4 Case Analysis: Successful Practices of Accessible Design

### 2.4.1 Accessible Product Design

Centaur is a revolutionary electric wheelchair. It aims to end social isolation from mobility issues. With a small size and great maneuverability, it allows users to access tight spaces and sit at dining tables. Its innovative software ensures safety and ease of use (see Fig. 8).

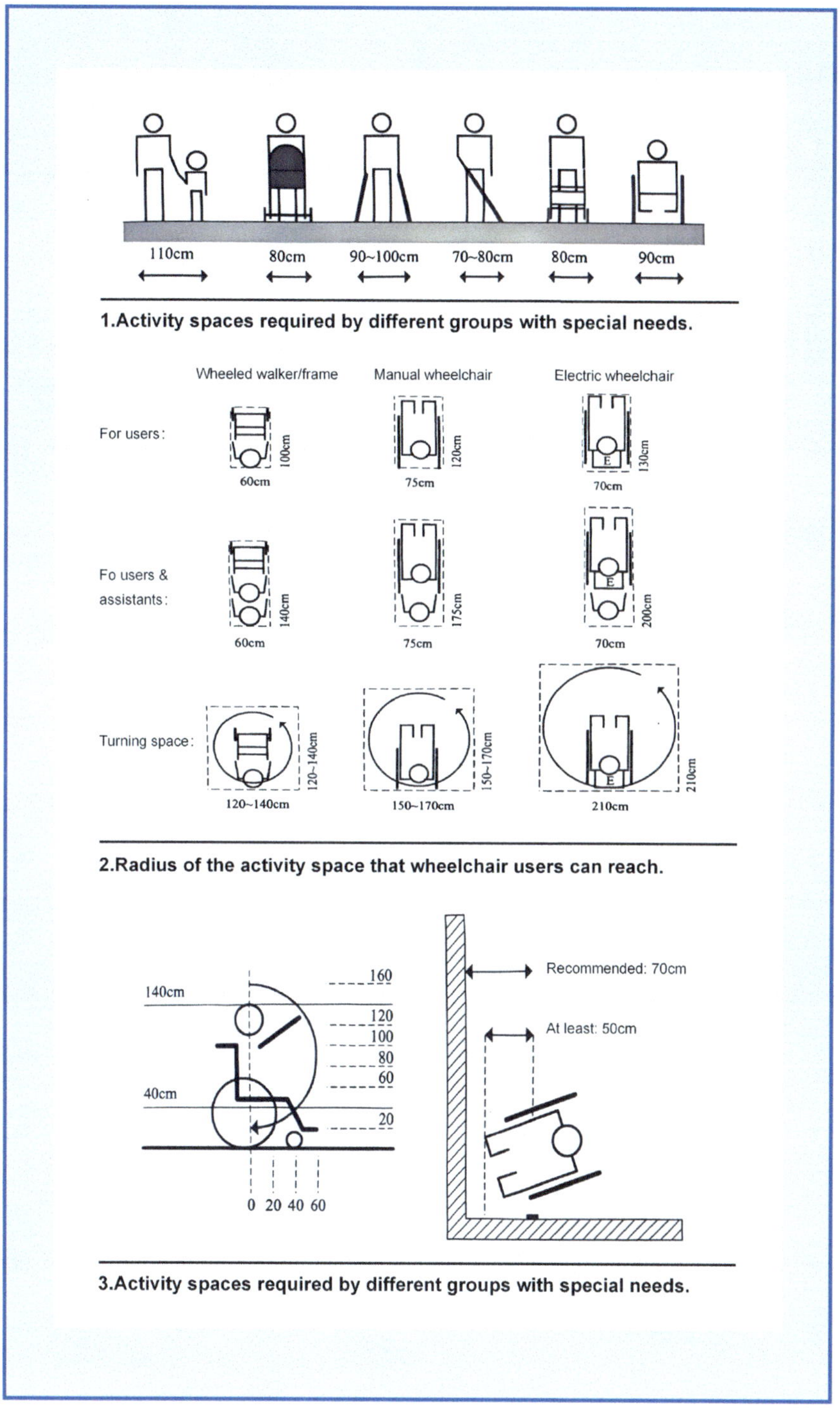


**Fig. 6** Accessible design dimension requirements for different groups. *Image source* Design Guide for Elderly-Friendly Residences

**Fig. 7** Intersection of accessible design and age-friendly clothing

**Fig. 8** Accessible product design

### 2.4.2 Accessible Environment Design

Its compact size and excellent maneuverability create new possibilities. Users can navigate narrow doorways, sit at the dining table, and access higher areas. The innovative software ensures safety and ease of use (see Fig. 9).

Home environment accessible design is crucial. It boosts the elderly's independent living and cuts nursing reliance, enhancing family well-being.

Accessible home bathrooms need enough turning space for wheelchairs, handrails, suitable toilets and showers. Accessible kitchens require adjustable workbenches, easy-reach cabinets/ appliances, and wide passages for mobility aids (see Fig. 9).

# 3 Principles of Safety and Usability

## 3.1 The Importance of Safety in Age-Friendly Design

Safety is vital in age-friendly design, safeguarding the elderly's health and life. As they age, declining physical functions make them more vulnerable to accidents like falls. So, age-friendly design must focus on safety. This means cutting physical barriers, adding anti-slip features, installing handrails, and ensuring stable environments (see Fig. 10).

Safety-centric design is essential everywhere. It protects the elderly's lives and boosts their lives and boosts their independence and confidence, allowing more autonomous daily living.

**Fig. 9** Accessible environment design

## 3.2 The Importance of Usability in Age-Friendly Design

Usability is key in age-friendly design, revolving around four main points. First, simplify operations for easy use by the elderly. Second, incorporate multi-sensory feedback to offset sensory decline. Third, use ergonomic s in product design for a better fit and less fatigue. Fourth, offer clear prompts for operation results and warnings. These work together to create user-friendly products, helping the elderly use items safely and comfortably and enhancing their quality of life and self-sufficiency.

## 3.3 Case Analysis: Successful Practices of Safety and Usability

### 3.3.1 EPT Safety Ergonomic Cane

Knee joint diseases are very common among the elderly, who often rely on assistive devices for movement and daily activities. For the elderly who

**Fig. 10** Safety design guidelines

have difficulty bending over, it can be extremely challenging to lift their legs onto the bed.

The EPT Safety Ergonomic Cane is designed to assist those suffering from knee joint diseases (such as inflammation, weakness, joint pain, etc.) with walking, getting into bed, and other daily activities without putting excessive pressure on the wrist (see Fig. 11).

(1) **Ergonomic handle**: Made of sweat-resistant medical-grade silicone. Adjustable in height to suit people of different statures.
(2) **Desktop balancer**: Crafted from recycled plastic. Complies with the principles of sustainable design.
(3) **Clip-type adjustment button**: Rotatable and height-adjustable to match the height of the desktop.
(4) **Form and Function**: It offers a hanging function. When pressure is applied to the handle, it can reduce wrist strain during long-term use. It enables easy walking, with the curved part rolling smoothly on the ground and the grippy sole at the bottom preventing slipping. The shape also serves as a hook, allowing users to lift or pull items closer without having to stand up from a sitting position. This reduces the risk of falls and other accidents, enhancing the quality of life and psychological sense of security (Fig. 12).

### 3.3.2 Clothes Clip-Type Portable Tracker

This portable tracker designed for the elderly is small, lightweight, and easy to clip onto clothing. It connects to a mobile app to provide real-time location information of the wearer, ensuring the safety of the elderly. Additionally, a safe zone can be set. When the wearer leaves the preset area, the system will issue an alarm. Moreover, in case of an emergency, the elderly can press a button on the device to send distress messages to pre-set contacts (see Fig. 13).

This tracker is especially suitable for the elderly with memory impairments or those prone to getting lost. It helps them maintain independence in daily activities while giving their families extra peace of mind.

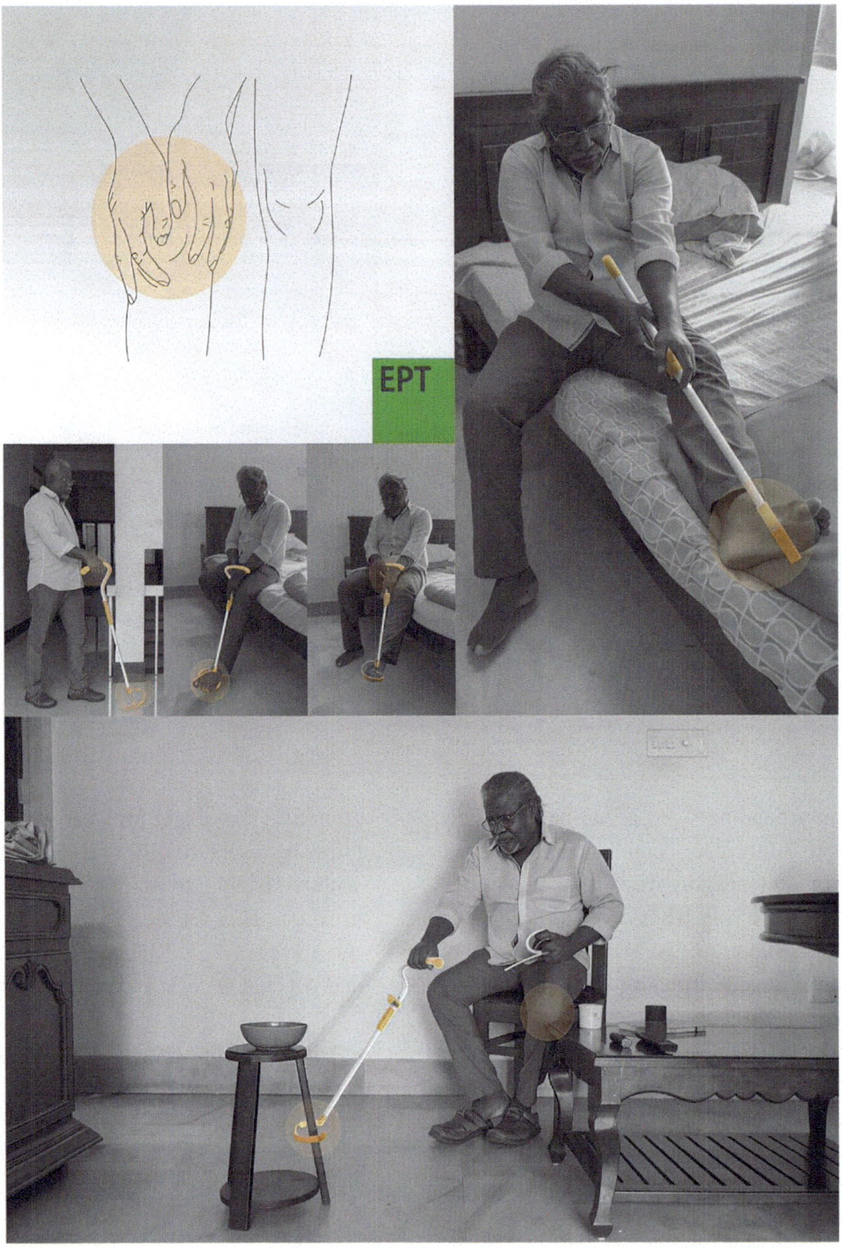


**Fig. 11** EPT tracker for elderly clothes. *Image source* https://www.sohu.com/a/481669997_121124036

# 4 Respectful and Inclusive Design

## 4.1 The Importance of Respect in Age-Friendly Design

In 1943, Abraham Maslow proposed Maslow's Hierarchy of Needs. It categorizes human needs into five levels: physiological, safety, social, esteem, and self-actualization.

When applied to the needs of the elderly population, we can clearly see the manifestation of needs at each level. Currently, an analysis of the demand situation of the elderly in China shows that the physiological and safety needs of the elderly have been basically met. However, the

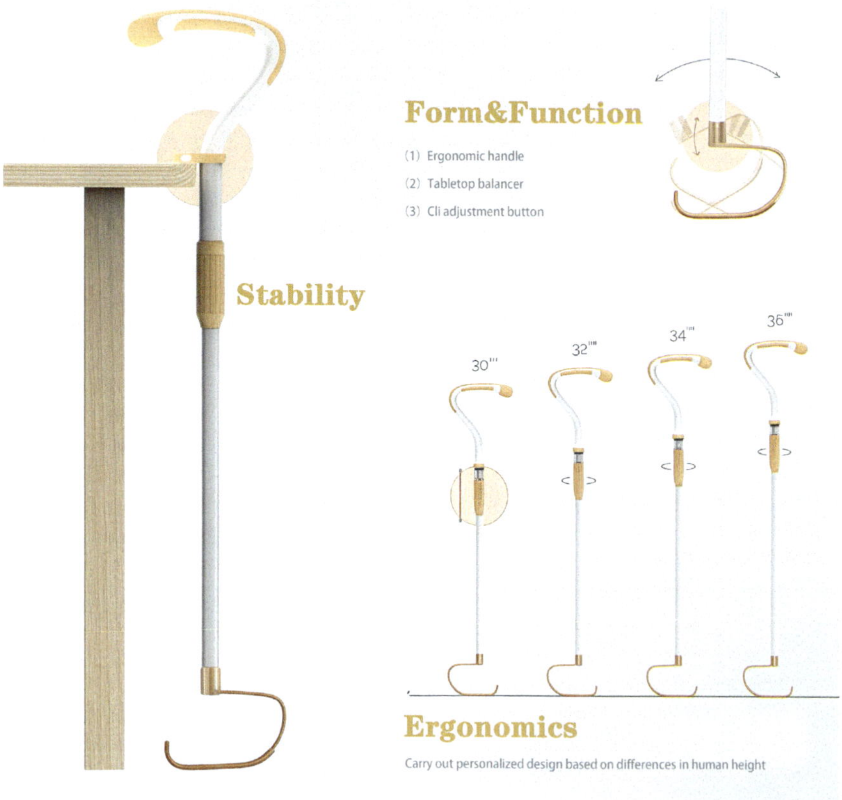


**Fig. 12** EPT tracker for elderly clothes

satisfaction of higher-level needs such as emotional needs, esteem needs, and self-actualization needs is still a cause for concern (as shown in Fig. 14).

As the elderly grow older, their labor ability declines and their social circle shrinks, making them prone to loneliness. Therefore, they long for social respect and attention, and hope to maintain dignity and autonomy in the community and family. They do not want to be regarded as complete dependents or incapacitated people. Even if they have limited mobility, they firmly believe that they can contribute to the family and society, and seek their own self-worth and the meaning of life. Thus, especially the element of respect, is of utmost importance in age-friendly design.

## 4.2 The Importance of Inclusivity in Age-Friendly Design

Inclusive design aims to offer equal access to a broad user base. It stresses considering diverse needs like age, gender, ability, and cultural background during design, ensuring products and environments are user-friendly for all.

For the elderly, inclusive design guarantees equal usage experiences. It provides safe, easy-to-use, and comfortable living setups and products. This supports their self-esteem and independence, lessening reliance on external aid and enhancing life quality and social involvement. It balances functionality with the elderly's psychological and emotional well-being during use.

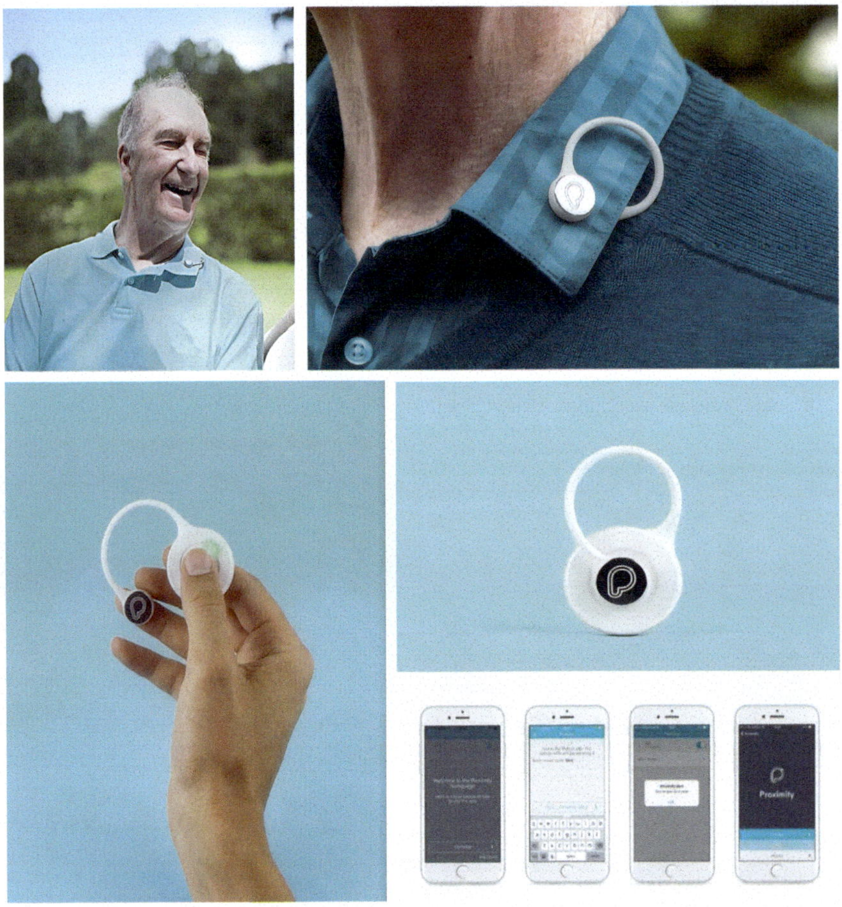


**Fig. 13** Boundary tracker for elderly clothes

The inclusive age-friendly design process builds a comprehensive framework. Key steps are as follows (as shown in Fig. 15).

## 4.3 How to Reflect Respect and Inclusivity in Design

### 4.3.1 User-Participation Autonomy

During the design process, ensuring user autonomy is a crucial way to reflect respect and inclusivity. By involving users in every stage of design, designers can understand their true needs and expectations, thus creating products or services that better meet user desires.

User feedback and opinions should be highly valued, preventing designers from unilaterally dominating the decision-making process. This approach makes design more personalized and user-friendly, and also makes users feel respected and valued during use, enhancing design acceptance and user experience (as shown in Fig. 16).

### 4.3.2 Diversity and Flexibility

Diversity and flexibility are core principles of inclusive design. They require design to adapt to the needs and preferences of different user groups. In practice, this means considering factors like users' gender, age, cultural background, and physical abilities, and offering multiple usage methods or options.

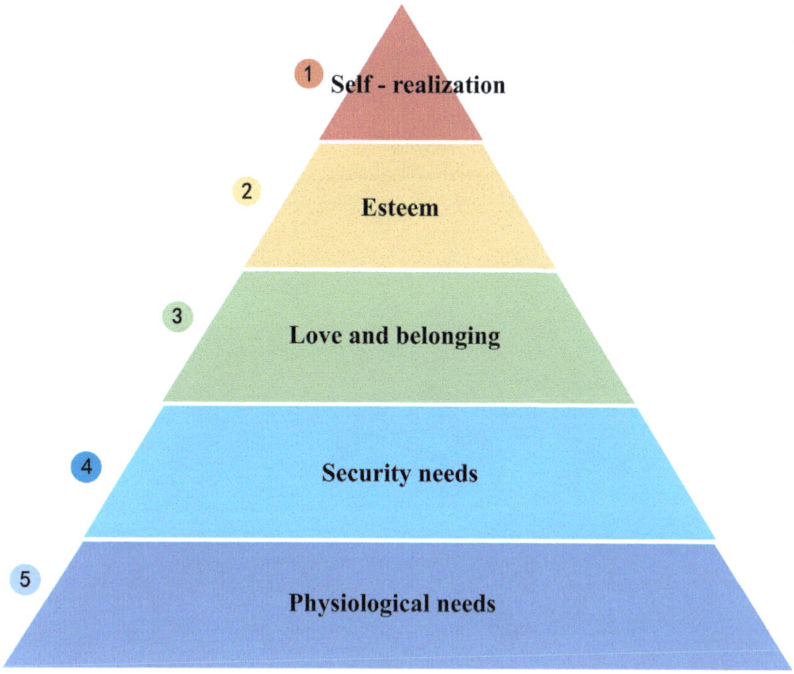


**Fig. 14** Hierarchy of needs model for the elderly

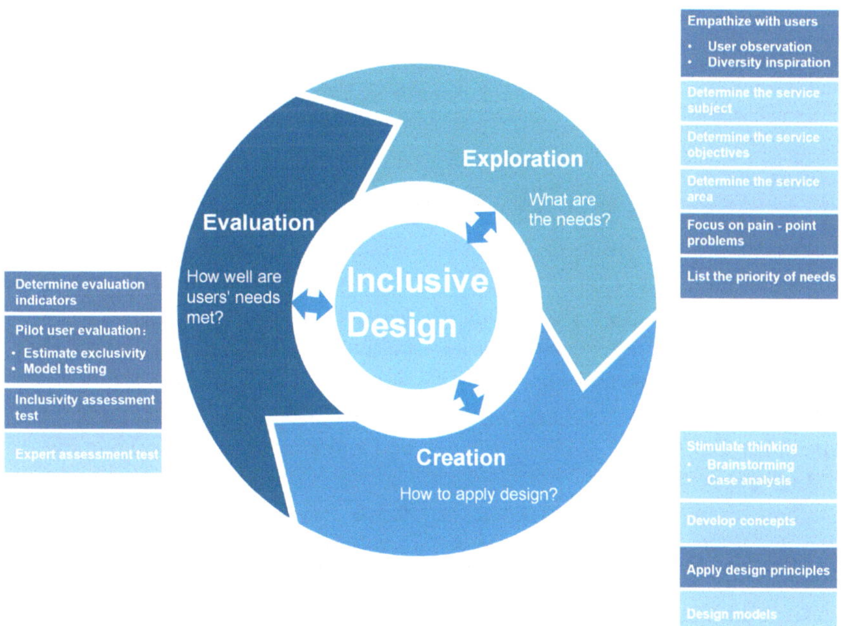


**Fig. 15** Inclusive design process

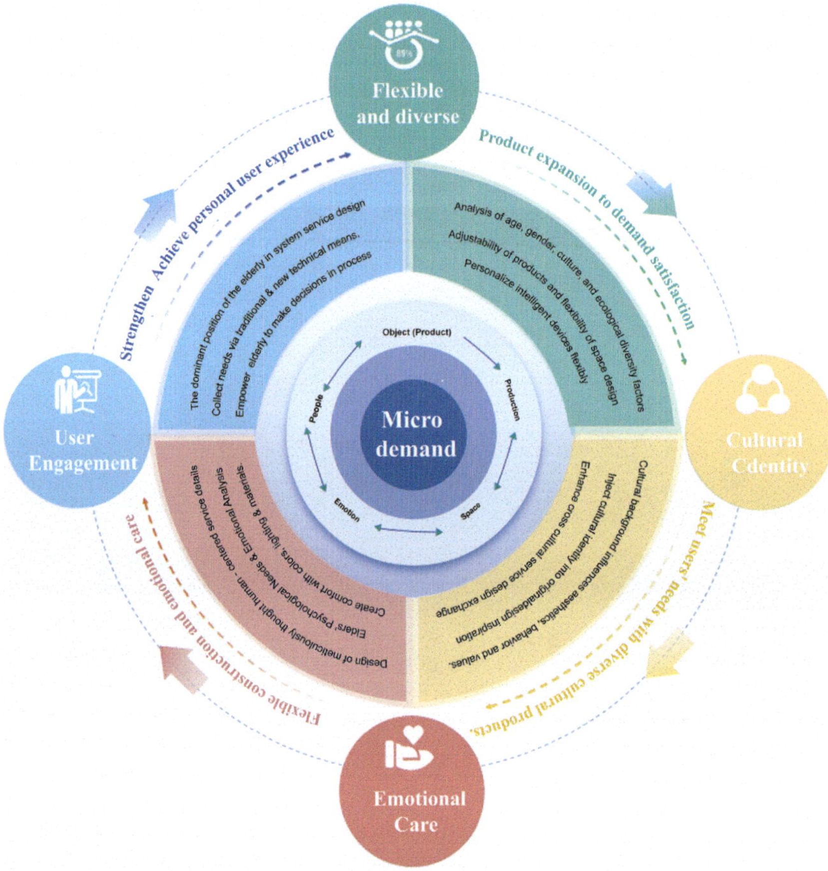


**Fig. 16** Respect and inclusive design

Flexibility is shown in the adjustability and adaptability of the design, ensuring products or services can meet users' personalized needs. These measures help the design better suit a wide range of users, achieving true inclusivity and universal usability (as shown in Fig. 16).

### 4.3.3 Respect for Culture and Background

Respecting users' culture and background is an important aspect of achieving inclusivity in design. Culture and background have a profound impact on users' aesthetics, behavior patterns, and values. Therefore, design must take these factors into account and avoid using elements that may cause misunderstandings or offense.

By deeply understanding the cultural and social backgrounds of target users, designers can create products or services that are closer to users' lives and needs, strengthening users' cultural identity and comfort. This not only increases the market acceptance of the design but also promotes cross-cultural communication and understanding (as shown in Fig. 16).

### 4.3.4 Psychological Comfort and Emotional Care

Psychological comfort and emotional care play a vital role in design, reflecting the importance attached to users' emotional needs. By designing products or services that meet users' psychological expectations and convey warmth and care, users can feel at ease and pleased during use.

This kind of design pays attention to details, such as gentle color choices, comfortable materials, and human-centered interface design. These can effectively reduce users' anxiety and stress when using products or services, thereby enhancing users' overall satisfaction and sense of happiness (as shown in Fig. 16).

# 4 Elderly User Research and Needs Analysis

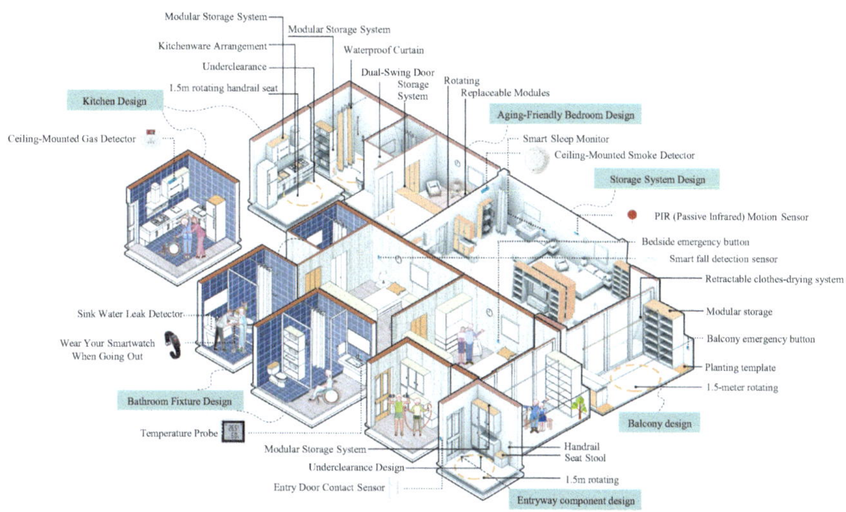


## 1 Defining Elderly Users

### 1.1 Defining Elderly Users

The core of design lies in being "human-centered," and user-centered design further emphasizes the need to fully consider user requirements at every stage of product development. In this process, defining the user serves as a pivotal step, aiming to gain an in-depth understanding of the characteristics and needs of the target audience. By analyzing demographic traits and user behavior, it becomes possible to identify the needs and pain points of different user groups, create user personas, and provide a critical foundation for design decisions. This ensures that the product effectively meets user expectations and enhances the overall user experience.

#### 1.1.1 User Persona

User persona design drives product iteration. By defining typical user behaviors, needs, and expectations (Fig. 1), it guide s design and service development. Established personas enable designers to predict user behavior, validate assumptions through testing, and refine solutions. Continuous feedback and testing ensure


W. Tang, *Aging-friendly Service System Design*, https://doi.org/10.1007/978-981-95-2054-1_4

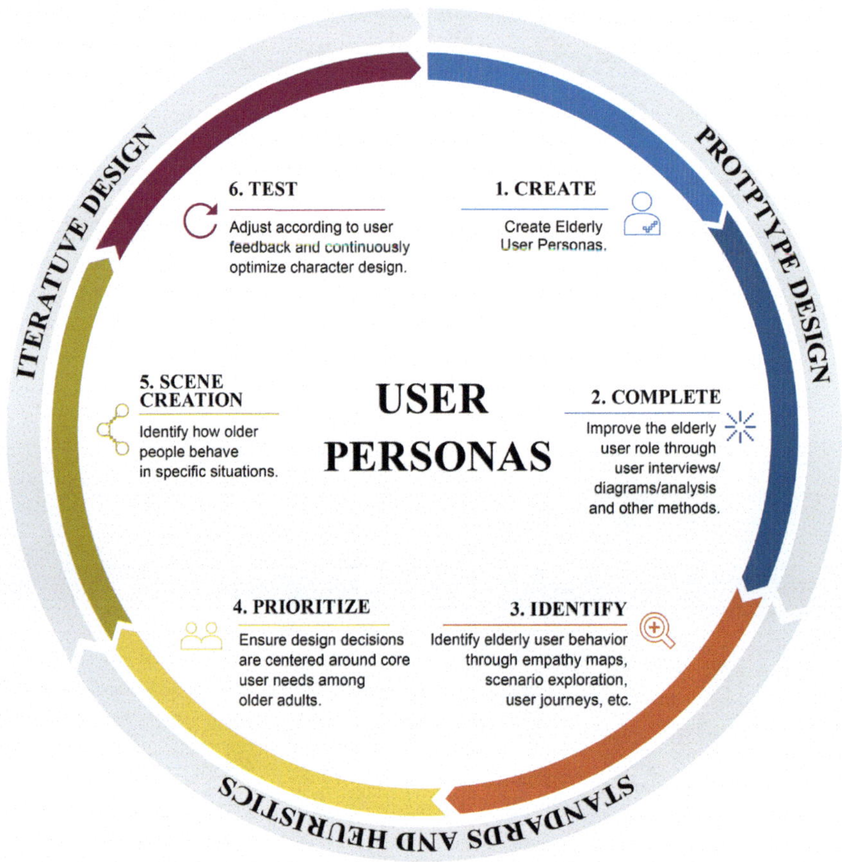


**Fig. 1** User role design

the final design meets user needs while improving functionality, usability, and safety. This approach is vital for goal alignment, issue refinement, and product relevance—ensuring practical, actionable outcomes.

### 1.1.2 Defining Elderly Users

The elderly user persona refers to individuals aged 60+. This group faces declining physical abilities, chronic conditions, and cognitive changes, impacting daily life and social engagement. Socioeconomic factors further influence their practical needs and emotional expectations. Defining this persona (Fig. 2) requires a holistic understanding of their characteristics and needs, incorporating multifaceted considerations to establish a robust foundation for research and design.

## 1.2 Research Methods and Tools for Elderly Users

Existing design research methodologies and tools can effectively support the entire design process when studying users. Below we introduce several common research approaches and instruments specifically adapted for elderly characteristics.

### 1.2.1 Qualitative Research Methods

Qualitative methods gather non-numeric, in-depth data on elderly users' experiences, interactions, and cultural contexts. Common methods include in-depth interviews, observations, and focus groups, capturing the complexity of elderly users' needs and experiences.

**Fig. 2** Defining the elderly user role

#### 1.2.2 Quantitative Research Methods

Quantitative methods collect precise numerical data through standardized tools, focusing on measuring relationships between variables and validating hypotheses. Common methods include surveys, experiments, and network analysis, providing insights into elderly users' behaviors and demands.

#### 1.2.3 Data Analysis Software

Data analysis software helps process and analyze data to extract meaningful insights. It includes tools for statistical analysis, qualitative data processing, data mining, and machine learning.

#### 1.2.4 Data Visualization Tools

Data visualization tools present data in visual formats (e.g., charts, dashboards) to simplify complex information. Common tools like Photoshop, Tableau, and Power BI enhance readability and usability of data.

## 2 Qualitative Research on Elderly Users

Qualitative research provides unparalleled depth in understanding elderly users' needs, behaviors, and experiences. By prioritizing direct engagement over numerical data, it captures the nuanced realities of aging—from daily challenges to unspoken expectations. This approach reveals what surveys cannot: the contextual why behind user actions, enabling truly human-centered design solutions.

### 2.1 In-Depth Interview Method

In-depth interviews are a highly valuable qualitative research tool, allowing designers to gain deep understanding of elderly users' personal experience s and insights. Through patient listening and thoughtful questioning, designers build rapport with elderly users. This process helps capture genuine needs, as well as understand their life backgrounds, mental states, and specific expectations for products or services. To ensure the interview's effectiveness, designers must use language familiar to elderly users, avoid technical jargon, and maintain a careful and attentive approach. Recording the interviews is essential for later analysis and the extraction of meaningful insights and recommendations.

### 2.2 Observational Research Method

The observational research method (Fig. 3) involves directly observing and recording

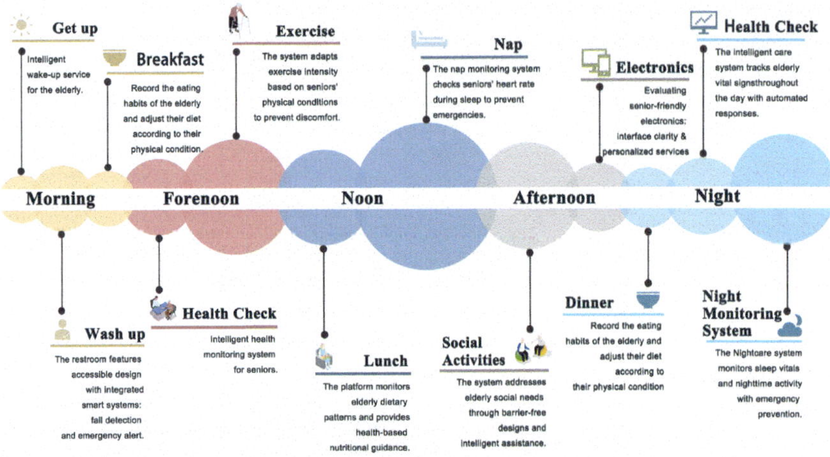


**Fig. 3** Schematic diagram of observational research method

behaviors, events, or situations to collect data. In elderly user research, this method is highly valuable. By observing elderly users' natural behavior s and subtle environmental interactions, designers can better understand their behavior and interaction patterns, as well as their responses to specific contexts. This method not only helps identify potential obstacles in product usage but also provides concrete, practical suggestions for design and service improvements. Designers must remain highly focused during observations and avoid interfering with the users' natural behaviors.

## 2.3 Focus Group Discussion Method

The focus group discussion method (see Fig. 4) is another important qualitative research tool that helps designers understand the collective views and opinions of elderly users. In group discussions, elderly users engage in conversations on specific topics, sharing their experience s and perspectives. This interaction stimulates more discussion and feedback, helping designers gain a comprehensive understanding of elderly users' needs and expectations. To ensure

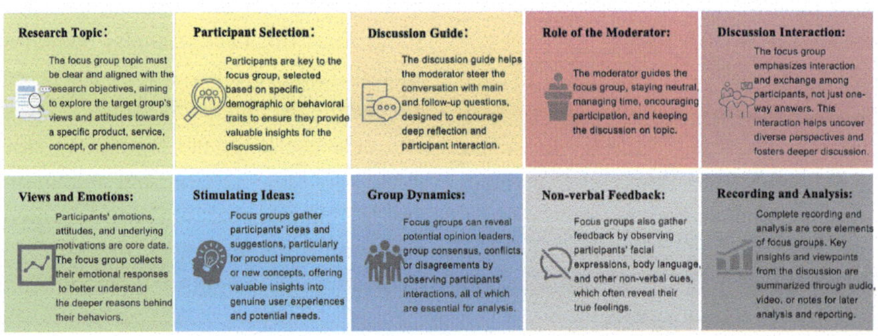


**Fig. 4** Core content of focus group discussion method

the effectiveness of focus groups, designers should prepare thoroughly in advance by selecting appropriate topics, participants, and creating a friendly atmosphere. During the discussion, managing time effectively and keeping the conversation focused is essential. Recording the session with video conferencing software (e.g., Zoom) or audio devices is crucial for preserving key information and insights, providing a solid theoretical and practical foundation for subsequent design work.

## 3 Quantitative Research on Elderly Users

Quantitative research on elderly users (Fig. 5) uses statistical analysis to understand and evaluate the needs, behaviors, and attitudes of the elderly population. This method relies on structured data collection tools, providing concrete and measurable results. Through surveys, experimental designs, and other tools, researchers collect large amounts of data that reveal trend s and characteristics of elderly users, offering strong support for policy-making and product design.

### 3.1 Survey Method

The survey method is a widely employed quantitative research approach, enabling systematic data collection on elderly users' demographics, usage patterns, and satisfaction levels. Online platforms (e.g., Google Forms, SurveyMonkey, Wenjuanxing) enhance efficiency by streamlining data acquisition, improving response rates, and ensuring robust quantitative outputs.

**Fig. 5** Common methods of quantitative research

### 3.2 Data Cleaning and Preprocessing

This phase ensures data quality through the removal of invalid, duplicate, or inconsistent entries. Standardization and normalization procedures align heterogeneous data for unified analysis, establishing a reliable foundation for subsequent examination of elderly users' behaviors and preferences.

### 3.3 Data Processing and Analysis

Statistical techniques (descriptive, inferential, and regression analyses) are applied via tools like SPSS or Excel to discern trends, test hypotheses, and derive actionable insights. Results are visualized (Fig. 6) to facilitate interpretation and support evidence-based design decisions.

## 4 Identification and Prioritization of Elderly Users' Needs

Elderly User Needs Identification and Prioritization Identifying and prioritizing elderly user needs is crucial for designing age-friendly service systems. In the early stages, designers utilize both qualitative and quantitative data to identify core needs and rank them based on their importance. This process ensures clear direction for design and guarantees that the final outcomes effectively meet elderly users' needs, providing a strong foundation for successful implementation.

### 4.1 Elderly User Needs

Through systematic research and analysis, designers build a clear framework of elderly user needs to guide the design of age-friendly service systems and enhance users' quality of

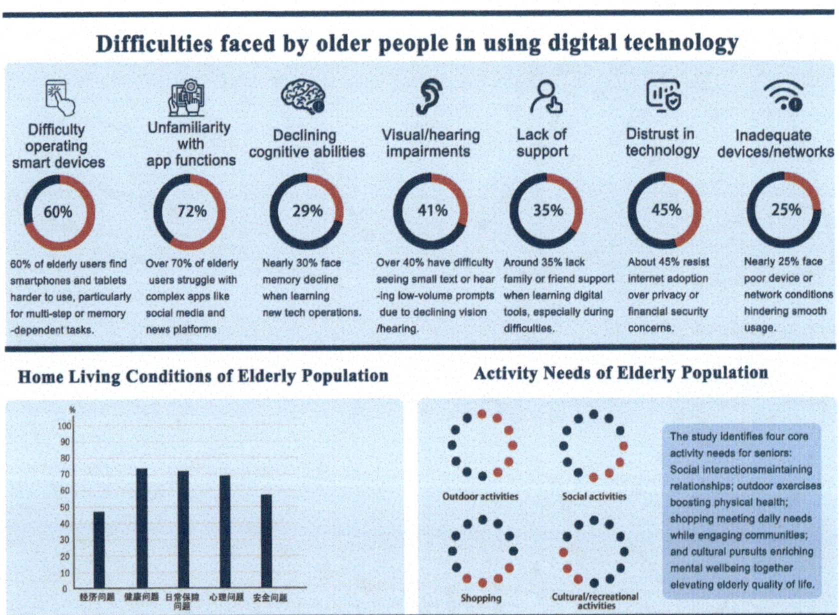


**Fig. 6** Schematic diagram of quantitative analysis data visualization

life. In the needs selection and organization phase, designers focus on key needs by eliminating redundancies and categorizing them according to their importance and urgency. This results in a structured, user-centered list of needs, providing clear guidance for design and development. In the needs analysis phase, designers identify key needs and common challenges, determining their priority based on urgency and impact, and suggesting actionable design improvements. During the needs exploration phase, designers delve deeper into both obvious and hidden needs, analyzing user behaviors and motivations to identify gaps in current products or services and propose targeted solutions. This comprehensive approach ensures a more thorough understanding of elderly users' real needs, offering practical and actionable design guidance.

## 4.2 Prioritization Methods and Applications

Scientific evaluation tools help rank the importance of elderly user needs, facilitating precise resource allocation and optimizing design solutions to meet core needs. Continuous feedback and adjustments improve the design's effectiveness, enhancing elderly users' quality of life and experience. The prioritization process considers factors like technical feasibility and resource allocation, focusing on needs that provide the most value and are easier to implement. Flexibility is key, as the process allows for adjustments based on feedback and changing needs. Safety, including environmental safety, health monitoring, and emergency systems, should be prioritized, followed by basic needs such as food, housing, and healthcare services. Psychological and social needs, including social activities and emotional support, should also be addressed to improve users' overall quality of life. Additionally, providing personalized and sustainable services that cater to individual differences is essential for long-term adaptability. After confirming the needs, methods like the KANO model (Fig. 7), MoSCoW analysis, and AHP can be applied to prioritize them. The KANO model helps prioritize needs based on their impact on satisfaction, while MoSCoW categorizes (Fig. 8) needs into must-have, should-have, could-have, and will-not have, guiding designers to identify key requirements. Once prioritized, design plans should be developed and adjusted based on user feedback to ensure alignment with evolving needs.

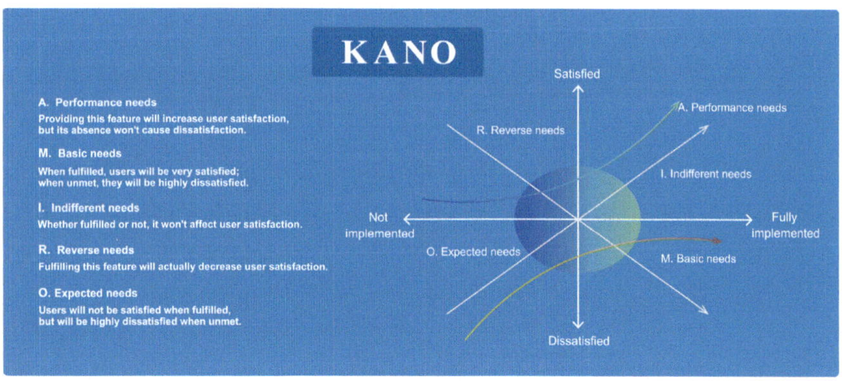


**Fig. 7** KANO model

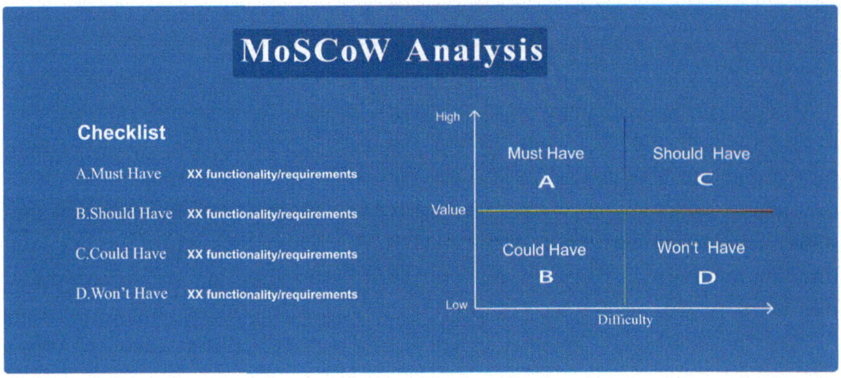


**Fig. 8** MoSCoW analysis method

# Interface Design of Elderly-Friendly Service System

# 5

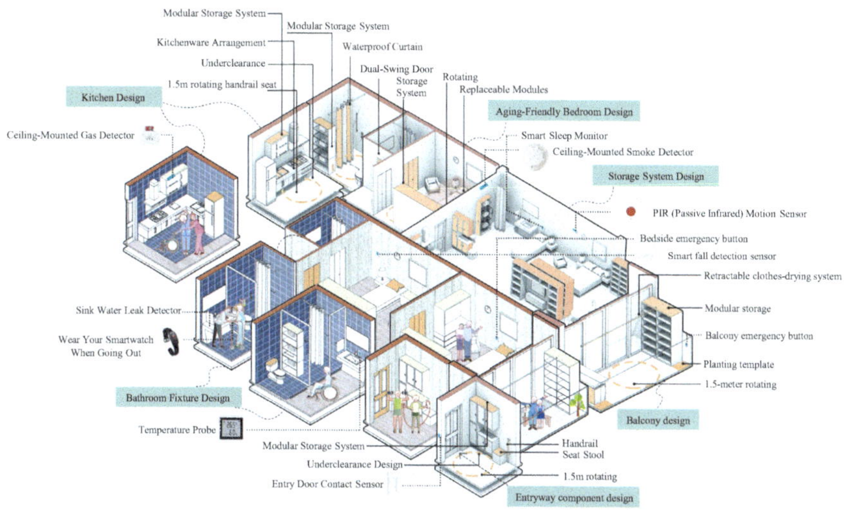


## 1 Design Style Suitable for Aging Service System

### 1.1 Design Style Suitable for Aging Service System Interface

In the interface design of aging service system, the mainstream interface styles in the market are skeuomorphic style, flat style and light mimicry style. These three do not exist in isolation, but gradually evolve with the development of design trends and changes in user needs, each with different advantages and limitations, so designers can often choose flexibly according to specific scenarios and user needs.

#### 1.1.1 Skeuomorphic Style

Skeuomorphism dates back to ancient Greece, when potters were already able to create metal—like handles, which became the earliest example of skeuomorphic design. Simulative style refers to the modeling and texture of real objects. Most of the ICONS use real objects in life to represent the actual functions of the product, highlighting details through superposition, highlight, shadow, texture, material, perspective and other effects, enhancing visual stimulation,


W. Tang, *Aging-friendly Service System Design*, https://doi.org/10.1007/978-981-95-2054-1_5

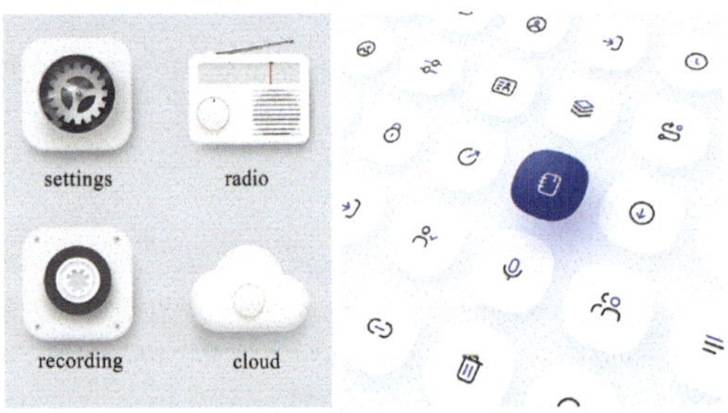


**Fig. 1** Skeuomorphic style ICONS

**Fig. 2** Flat style ICONS

helping users to easily get started and understand various functions at a glance (Fig. 1). By introducing familiar objects used by users in real life into the interface elements, it can help users easily judge reasonable and correct interface control methods based on their existing experience, greatly reducing the learning cost of users, reducing the use barriers of new users, and easing the fear of stranger s in operation.

Skeuomorphic style design has the advantages of easy recognition, easy to understand and easy to use, but there are also some problems. The cost of skeuomorphic design is high, it is difficult to update and iteration, the optimization space is limited, and the exquisite interface everywhere will lead to element crowding and easy to produce visual fatigue, which cannot meet the usage habits of most people. In addition, some designers pay too much attention to the details of the icon skeuomorphism, trying to completely imitate the real objects from the appearance, paying attention to the restoration of texture and texture, but the interface is not completely consistent with the actual after all, and some design working methods are still different from the actual function, or even very different, so there is a certain sense of separation. Therefore, in the skeuomorphic style design should pay attention to the choice, the details are too complex or too simple will easily confuse the user, it is difficult to identify the designer wants to express the content, resulting in the situation of not understand, but also try to meet the user's pursuit of quality, choose the appropriate material and texture. Different elements can bring a different look and feel, such as metal elements can produce a sense of science and technology, wood and leather elements can create a sense of time, natural elements can produce a sense of affinity, and reasonable planning to add appropriate light and shadow and color, enhance the three-dimensional effect of the icon, can bring a better visual experience (Fig. 2).

### 1.1.2 Flat Design

If skeuomorphism is to restore the three-dimensional texture of an object, then flatness is a two-dimensional visual representation. The flat style ignores light and shadow, texture and perspective, and forms simple graphics through lines and surfaces, which is more abstract, concise and more comfortable in visual effect than the delicate skeuomorphism (see Fig. 3). Flat style abstraction has disadvantages of recognition and cognition for elderly users. The elderly have limited cognitive ability, and it is difficult to recognize ICONS and components in flat style.

The flat design concept was proposed by Google in 2008. The core meaning of the concept of flatness is to remove redundant, heavy and complex decorative effects. This is manifested in the removal of superfluous perspective, texture, gradients, and elements that can make 3D effects, so that the "information" itself can be highlighted as the core. At the same time, in the design elements, it emphasizes abstraction, minimalism and symbolization. Excellent interface interaction design should not only realize the simple form, but also ensure the technical simplicity, the pursuit of less is more, the simple to control the complex, the focus is

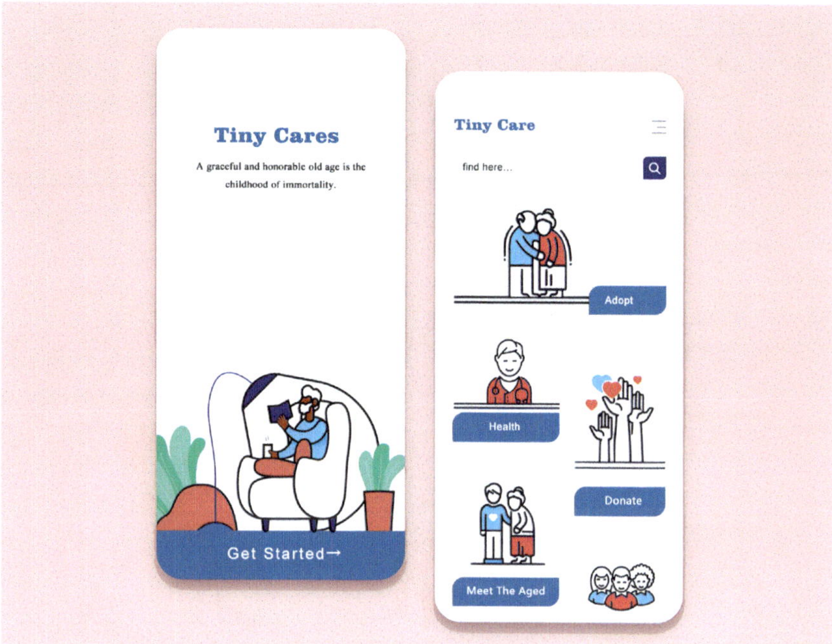


**Fig. 3** Flat style

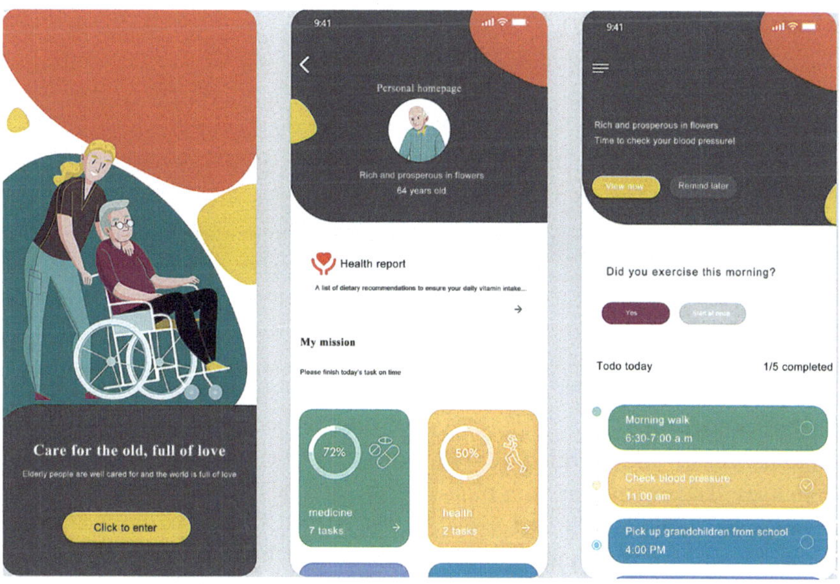


**Fig. 4** Flat style interface continuation image

highlighted and maintain an internal harmony. After all, color accounts for more than 80% of the initial visual experience of viewing an object (Fig. 4). In addition, it is also necessary to accurately grasp the emotional color contained in the color, fit the emotional cognition of the user, and then better mobilize the emotion of the user (Fig. 5).

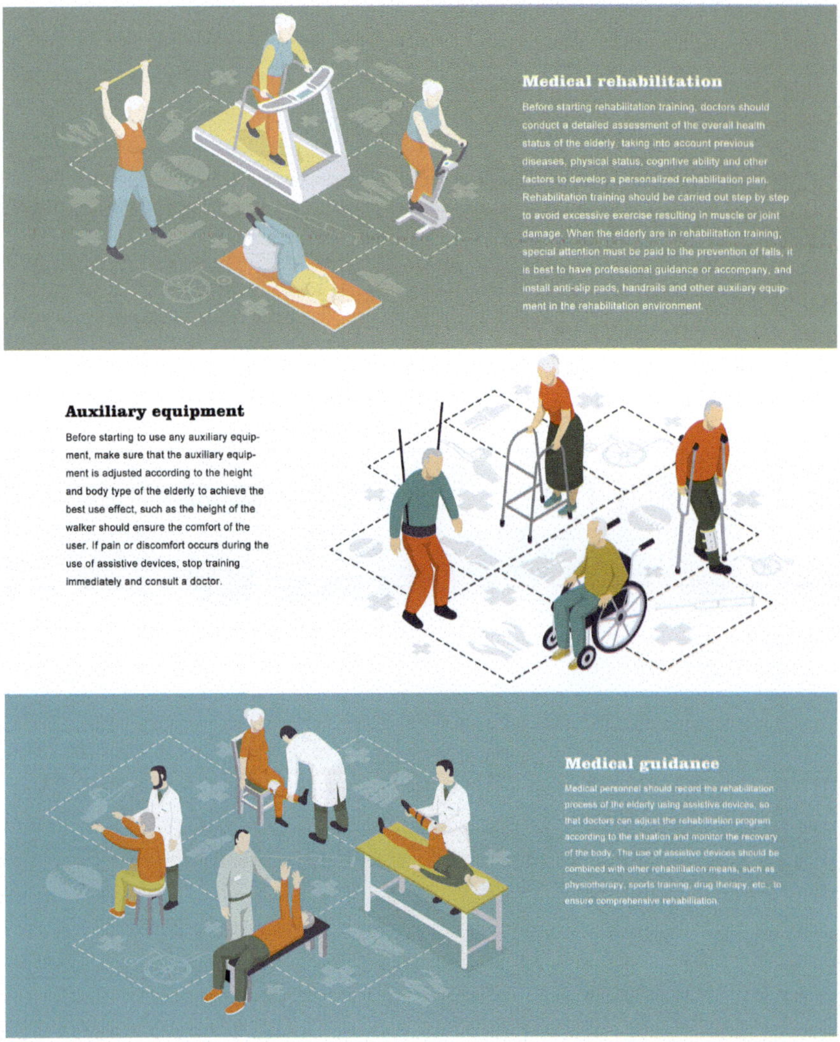


**Fig. 5** Flat style interface

### 1.1.3 Soft Skeuomorphism

Soft Skeuomorphism is an emerging design style that combines skeuomorphism with minimalism, retaining the subtle texture and visual layering of skeuomorphism, but preferring simplicity and lightweight overall. While simplifying, this style retains some features of real objects, Fig. 4: Flat style interface making the interface design both realistic and modern (Fig. 6). Soft Skeuomorphism focuses on the concept of minimalist design, and the ICONS, buttons and other UI elements are removed from complex details while retaining a certain sense of three-dimensional and texture, making them appear clean and modern. Emphasize gentleness in visual effects and avoid strong contrast effects. Compared with the new mimicry design, the light mimicry shadows and highlights are softer and slighter, presenting a more soothing visual experience overall (Fig. 7). Soft Skeuomorphism creates a suspended effect through highlights and shadows, which is more streamlined than the flat style. The light mimicry style has appeared in the Huawei HarmonyOS UX design concept promotional video. The simple atmosphere created by the

**Fig. 6** New mimetic style interface continuation image

light mimicry style and the sense of space left blank are favored by many young people, and the interface style with more information elements in the past has formed a strong contrast, and some fashionable apps have also been revised. However, because it is too simple, the interaction of the operation is not obvious, difficult to popularize, and not suitable for most users. For elderly users, its disadvantages are more obvious, including unclear functionality, fuzzy operation direction, color contrast is too soft, so the current light mimicry style can not meet the needs of elderly users (Table 1).

## 2 Changes in Perceptual and Cognitive Abilities

In interface design, the choice of fonts, backgrounds and ICONS should be carefully considered to ensure the overall visual balance. In the selection of fonts, in order to ensure the readability of text and the reasonable presentation of dynamic effects, it is necessary to pay attention to its size, color and animation characteristics. Background design should take into account the unity of color, texture and form to ensure that the interface has a visual harmony of

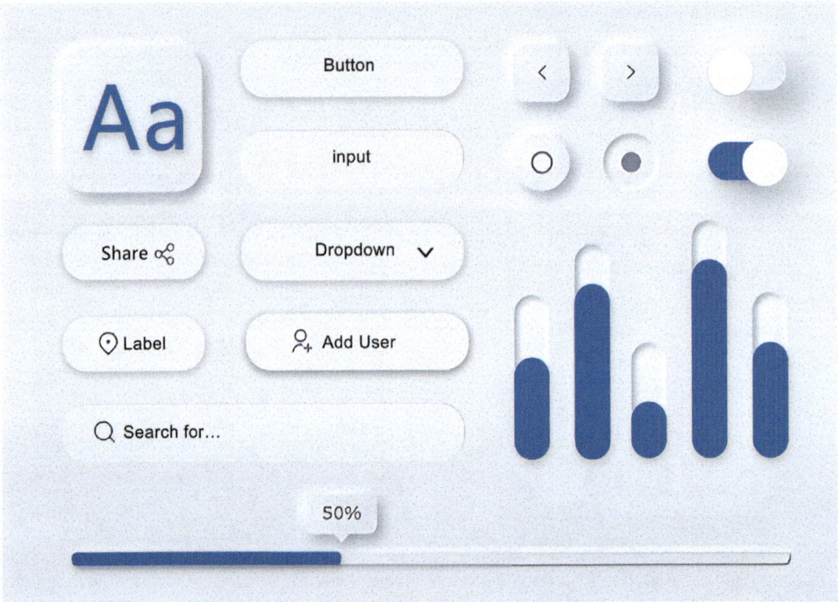


**Fig. 7** New mimetic style interface

**Table 1** Advantages and disadvantages of the three styles

| Design style suitable for aging service system interface | | | | |
|---|---|---|---|---|
| English name | | Peculiarity | Advantage | Disadvantage |
| 1 | Skeuomorphism | Simulate the shape and texture of real objects by superimposing high-lights, textures, materials, shadow and other effects | Low cognitive and learning costs | Can limit the design of the function itself |
| 2 | Flat design | Remove redundant, heavy decorative effects and use a more straightforward design to get the job done | Highlight the theme of the content, simple but not simple | Requires a certain cost of learning, to convey feelings is not rich |
| 3 | Soft Skeuominimalism | It retains the subtle texture and visual hierarchy of the skeuomorphic design, but tends to be concise and lightweight on the whole | Shadows and highlights are softer, resulting in a more soothing visual experience overall | The interactivity of the operation is not obvious |

beauty, while providing effective support for the content. The design of ICONS needs to strike a balance between simplicity, straightness and accessibility, so that they can not only convey information quickly, but also have a high degree of recognition (Fig. 8).

## 2.1 Service System Interface Font Design

The structure of the human eye is like a precision camera, the process of seeing depends on the light reflected to the surface of the object, through the refractive media such as cornea,

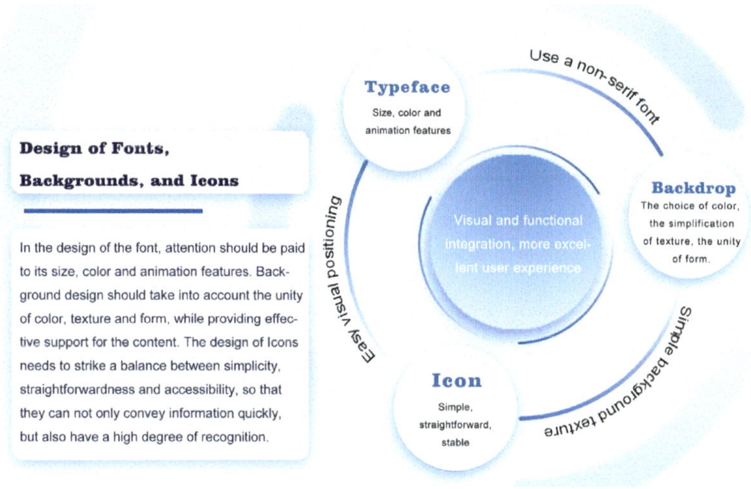


**Fig. 8** Selection of font, background and icon

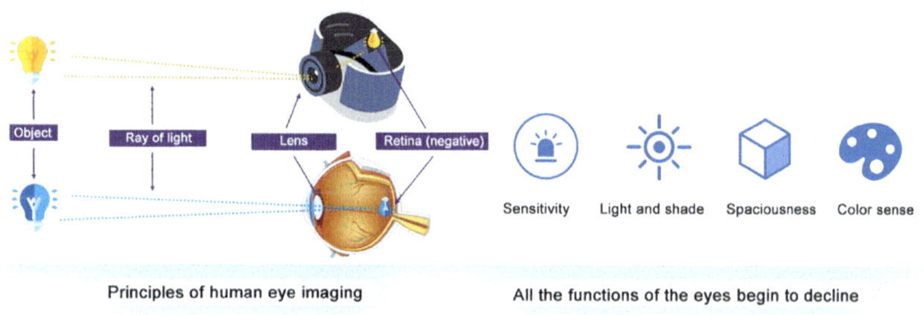


**Fig. 9** Why can't the elderly see clearly

lens and vitreous refraction, and finally form an image on the retina. However, with age, the lens gradually hardens and the function of the ciliary muscle weakens, causing objects to be unable to focus clearly on the retina (Fig. 9). Studies show that at age 40, only about 50% of light reaches the retina as effectively as it does at age 20, and by age 60, that number drops to 20%. The above physiological degradation directly leads to a significant decline in the acuity, light and dark perception, space perception and color perception of the elderly in visual discrimination of objects, which leads to the problem of blurred vision.

Designers should avoid reducing age-friendly design to a simple "scale-up" solution—that is, simply increasing the font, volume, or screen size to address the needs of older users (Fig. 10). This too one-sided treatment ignores the compatibility of the overall interface, may lead to the layout of the page elements, and even the dislocation of the verification code, which increases the difficulty of use, but brings unnecessary trouble to the user. Age-appropriate design should pay attention to the overall optimization, provide a truly effective and compatible design scheme through in-depth insight into the actual use situation of elderly users, balance the multi—sensory experience such as vision and hearing, and enhance the interactive experience and convenience of users.

In interface design, the setting of font symbol size is not only related to the readability and visual hierarchy of information, but also needs to consider the needs of different user groups. The selection of colors should comprehensively consider readability, contrast and emotional

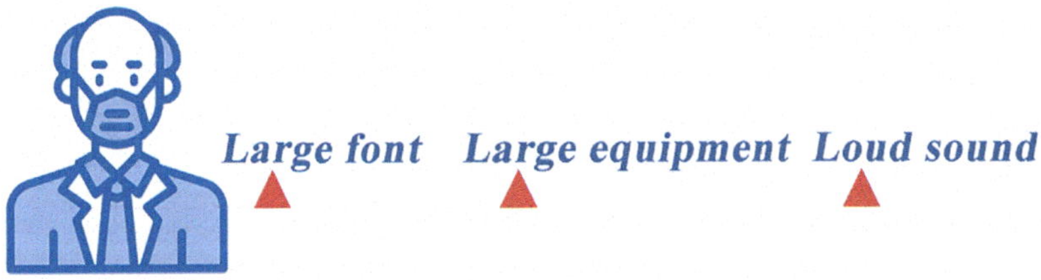


**Fig. 10** Age-friendly design should not only consider "big"

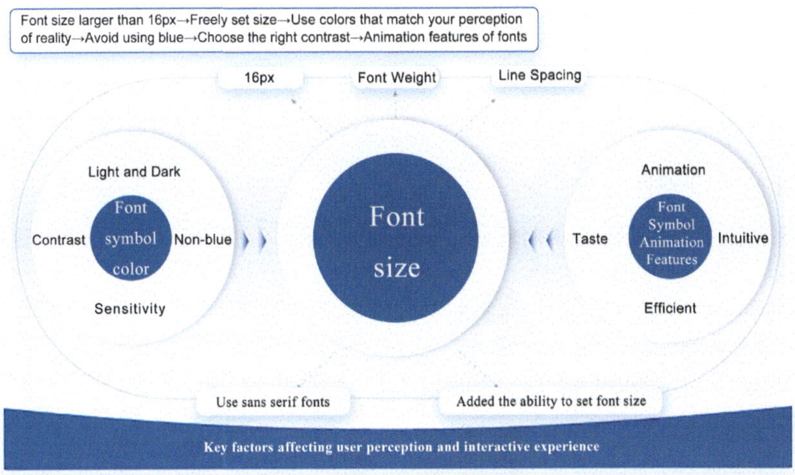


**Fig. 11** Key Factors Influencing User Perception and Interaction Experience

communication to ensure that the recognition of information can be effectively improved in diverse scenes. As for the animation features of font symbols, their application should follow the principle of smooth and natural, which should not only enhance the visual guidance of users, but also avoid the interference caused by excessive dynamic effects to ensure the comfort of user experience (Fig. 11).

In the selection of fonts, especially for the design needs of elderly users, fonts are generally divided into serif and non-serif. Serif bodies, the so-called "footers", refer to typefaces with edges and corners, while non-serif bodies are characterized by mechanization, uniform line thickness, and no corner decoration (Fig. 12). Based on visual perception, non-serif bodies often have larger glyphs and a more concise and clear structure, which improves readability in long text reading and small screen displays. Therefore, non-serif bodies are more suitable for situations where there is a lot of text or limited screen space. Serif, on the other hand, can enhance the beauty of short sentences due to its elegant decorative edges and corners, and is often used as the preferred font for titles or short texts to enhance visual appeal and aesthetic effect.

### 2.1.1 Font Size

With age, the field of vision of the elderly gradually shrinks, so when designing an interface with a large amount of text, it is necessary to pay special attention to the following key points.

**Fig. 12** The difference between serif bodies and sans-serif bodies

## 2.2 The Text Size Is Not Less Than 16 Pixels

Although studies have shown that 12-pixel fonts are adequate for normal reading, they are still too small for most users, especially the elderly. Even a 15-pixel font can cause potential users to stop browsing because of reading difficulties. So why is it recommended to set the text size to 16 pixels? The study found that 16-pixel text appears on the screen in a size similar to text in printed matter (such as books and magazines), which fits people's reading habits. In addition, when screen reading, the distance between the user and the screen is an important factor that cannot be ignored. Therefore, the 16-pixel font, while maintaining a comfortable reading experience, can also better adapt to the distance between the screen and the reader, ensuring clear and efficient information transmission (Fig. 13).

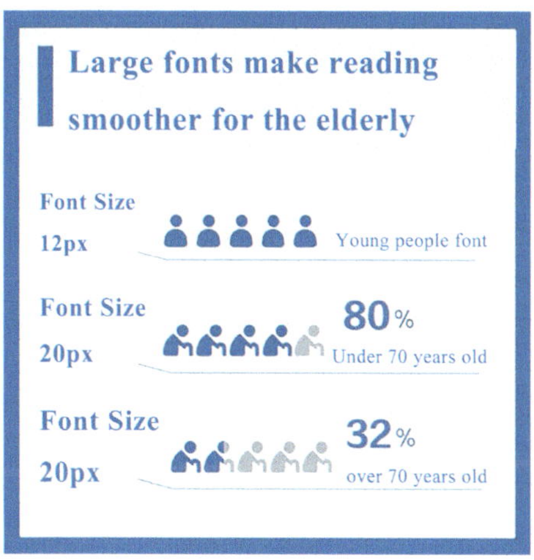


**Fig. 13** Large fonts make reading smoother for the elderly

Take reading books as an example, the printing font of books is generally set between 10 and 12 o' clock. In the digital scene, the 16-pixel screen font can be visually matched with the printed font, so that the reading experience of the two is matched. Everything that involves reading or clicking should be enlarged according to the appropriate proportions, especially when designing the text on the button, the font should remain large and should not be smaller than 16 pixels (Fig. 14).

After calculation, it is determined that the minimum size of the font size is 16 pixels. In product design, common word weights are Regular and Medium (Fig. 15). However, for aging products, in order to further improve the readability of the text, it is recommended to add an additional word weight on the basis of these two words, and make the text outline more sharp by bolding the text. This approach can effectively enhance the clarity of the text and optimize the reading experience of elderly users.

## 2.3 Add the Ability to Set the Font Size

In the case of limited screen resolution, usually the title line spacing is 1.3 times, and the text line spacing is 1.5–1.8 times, and the visual effect is the most comfortable. However, a unified standard scheme may not be able to meet the needs of all users, and the simple pursuit of large fonts may reduce reading efficiency. Therefore, it is recommended to provide users with the option to customize the font size in the Settings, and provide a variety of sizes to adapt to different reading needs and time periods.

**Fig. 14** Requirements for font size in book reading

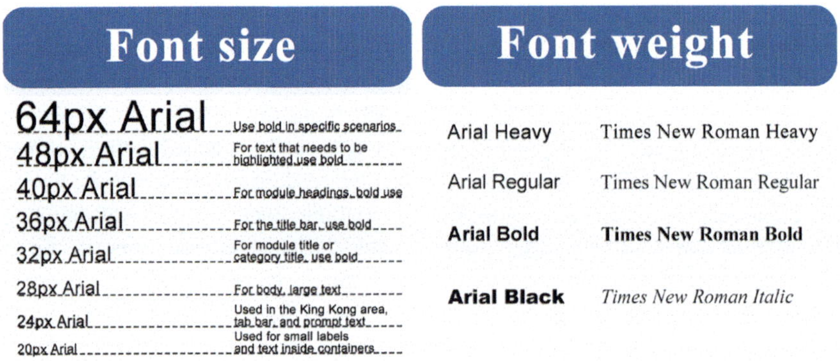


**Fig. 15** Schematic diagram of font size and weight

Designers should adjust the font size step by step according to the content hierarchy, such as title, body, and comments, or use different font sizes according to the importance of the content, to enhance the recognition of the text and the hierarchy of the design.

### 2.3.1 Font Symbol Color

In the interface design, the color selection of font symbols should avoid the mixing of too many colors, so as not to cause visual confusion and information disorder. The color that forms a clear contrast with the background color should be preferred to make the text more visually prominent, thus improving the legibility of the text and ensuring the clarity and effectiveness of information transmission.

## 2.4 Use Colors That Match the Perception of Reality

Older people often rely on existing experience to understand things, especially in the context of Chinese culture, which is more inclined to think through experience than foreigners. Therefore, in the design process, the ease of use and intuitiveness of the design should be optimized as much as possible to reduce the learning cost of the elderly group. For example, in daily life, traffic lights at intersections are green for go, yellow for wait, and red for stop. The meaning conveyed by this color has been deeply rooted in the user's cognitive system. This principle can be used for reference in the design of important prompts, and the use of color can be consistent with the elderly's cognition of color in real life (see Fig. 16).

## 2.5 Avoid Using Blue Fonts

In order to alleviate the visual burden caused by long text, the use of pure black fonts should be avoided, and it is recommended to choose soft tones such as dark gray, which can effectively reduce visual pressure and make the text more light and smooth. At the same time, pay

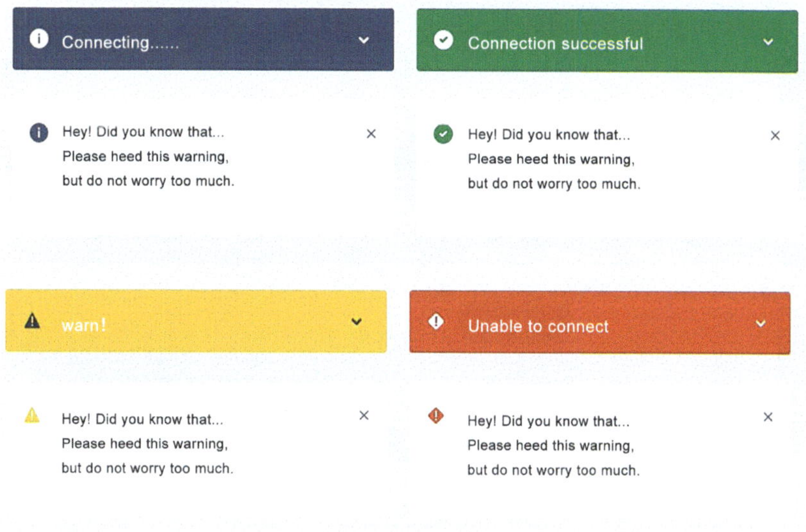


**Fig. 16** Prompt boxes of different colors

attention to try not to use blue, with the increase of age, the sensitivity of the elderly to color contrast decreased significantly, the difference between similar colors became worse, especially in the discrimination between blue and purple is more difficult, the recognition of green is also significantly weakened. Since the visual system of the elderly is slow to perceive blue shades, important elements of the interface should reduce the use of blue.

## 2.6 Select the Most Appropriate Contrast

In interface interaction, users need to quickly identify each element of the interface in order to perform correct operations. By using high-contrast colors, interactive elements such as buttons and ICONS can be made more visible, helping users easily identify which elements are clickable, which helps reduce misactions due to visual confusion, thereby improving the accuracy of operations. For older users, this design makes it easier for them to find the information and features they need when browsing the interface, thus completing operations more efficiently and reducing the time cost of searching and finding.

In addition, low-contrast and blurry interfaces can cause eye strain, while high-contrast color schemes can effectively alleviate eye discomfort for elderly users by enhancing the legibility of text and images (Fig. 17). In the "Web Content Accessibility Guide (WCAG) 2.0" mentioned the color accessibility design standard AA, small text and background contrast should be at least 4.5:1; Large text (bolded 14pt/normal 18pt and above) should have at least 3:1 contrast ratio (Fig. 18).

**Fig. 17** Contrast color ring

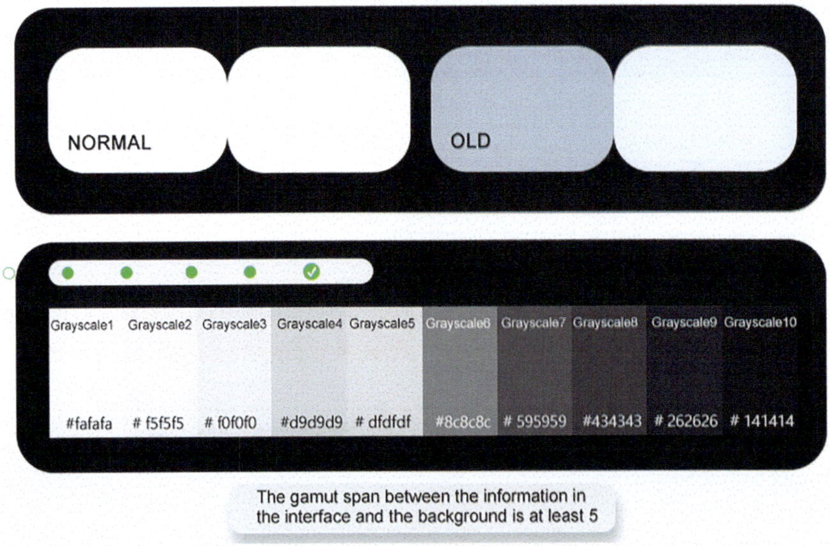


**Fig. 18** Interface information and background color gamut span requirements

While some users need high contrast, others (such as those with dyslexia) are unreadable for high—brightness colors and need low contrast. In order to meet such special groups, the design should include customizable color theme options. This means that the website or app should provide a user-friendly interface that allows users to adjust the color contrast of text colors, background colors, and elements such as links and buttons according to their personal preferences. By providing a variety of preset color schemes or allowing users to choose their own color codes, you can ensure that each user finds the best visual presentation for them.

### 2.6.1 Animation Features of Font Symbols

With the rapid development of science and technology, new media continue to emerge, and the "animation" of font symbols has also emerged. Compared with static text or pictures, animated font symbols can express information more vividly and intuitively, convey emotions and enhance narrative through visual effects. Good dynamic effect and visual design are complementary. In view of the confusion that aging groups may encounter in the process of use, animated design of font symbols can be introduced to express the information that static visual effects cannot accurately convey, such as feedback, guidance of the next step, state expression, etc. This design is not only more attractive to older users, but also can improve their understanding of the information, to achieve the perfect combination of interpretation, viewing and fun.

## 2.7 Service System Interface Background Design

In the background design of service system interface, the choice of background color, the simplification of texture and the unity of form together constitute the important elements to improve user experience and interface functionality, which can enhance the fluency of operation and the professional sense of the system. These design strategies, especially for older users and special needs groups, can effectively improve the usability and accessibility of service systems, ensuring that their wider user base can operate easily and comfortably.

### 2.7.1 Use Complementary Colors as the Interface Background

In digital interface design, complementary colors, as the most intense contrast of color combinations, can visually form obvious conflict and contrast effects, making elements more eye-catching and prominent. The background color and font color are complementary colors, which can effectively enhance the visual contrast in the interface and make the font more clearly visible. Compared with low saturation colors such as the Morandi colors popular in recent years, bright colors and strong contrasts are easier to attract the attention of the elderly group, and are also more practical in practical operation.

Common complementary color combinations are red and green, blue and orange, yellow and purple, etc., but in digital interface design, complementary colors with higher purity and brightness are often difficult to use directly, because they may appear too harsh and do not meet aesthetic requirements. Usually, designers will choose one color as the main tone, supplemented by a small area of complementary colors to enhance the visual texture of the interface. In addition, by reducing the purity and brightness of the color, or introducing neutral colors such as black, white, and gray to harmonize, the sense of conflict between complementary colors can be effectively reduced, thus making the interface more harmonious and unified.

### 2.7.2 Simplify the Background Texture

The simplicity of background texture is also important. A good background texture design should follow the principle of simplicity without losing depth. It should not only enhance the readability and practicality of the interface, but also provide strong support for information transmission. It is also necessary to build visual layers, give the interface a sense of space, and try not to attract the attention of the user, so as to promote the immersion of the user. The simple background texture layout that is in line with the reading habits and memory preferences of the aging group can effectively prevent the potential interference of diverse topics on users' thinking and ensure the coherent absorption of information.

### 2.7.3 Unify the Background Form

In view of the stability of the operation habits of the elderly group, the design should keep the same function or the same interface, so that the elderly can more easily integrate and extract the required information according to the visual memory clues, so as to make the operation process more smooth and pleasant. At the same time, providing a consistent feedback mechanism for the same type of operations can not only enhance the operational familiarity of elderly users, promote them to grasp and adapt to the interface style and interaction logic more quickly, but also significantly improve the overall user experience and make the interface operation more friendly (Fig. 19).

## 2.8 Service System Interface Icon Design

ICONS are one of the essential elements of interface design. Clear and intuitive ICONS can clearly refer to the meaning, facilitate visual positioning, and strengthen recognition memory. In the icon design of the interface of the service system, the minimalist design emphasizes the expression of the richest information with the least elements, and removes redundant decorations through refined lines and color matching, so that the icon looks clean and crisp, which is easy for users to identify quickly. Direct design directly shows the core meaning of the icon, avoids the use of complex or obscure symbols, and ensures that the user can understand the function or state of the icon at a glance. Stable emphasizes the continuous consistency of the interface and aims to reduce the learning burden and adaptation barriers caused by interface changes, thus ensuring that older users can easily operate digital devices.

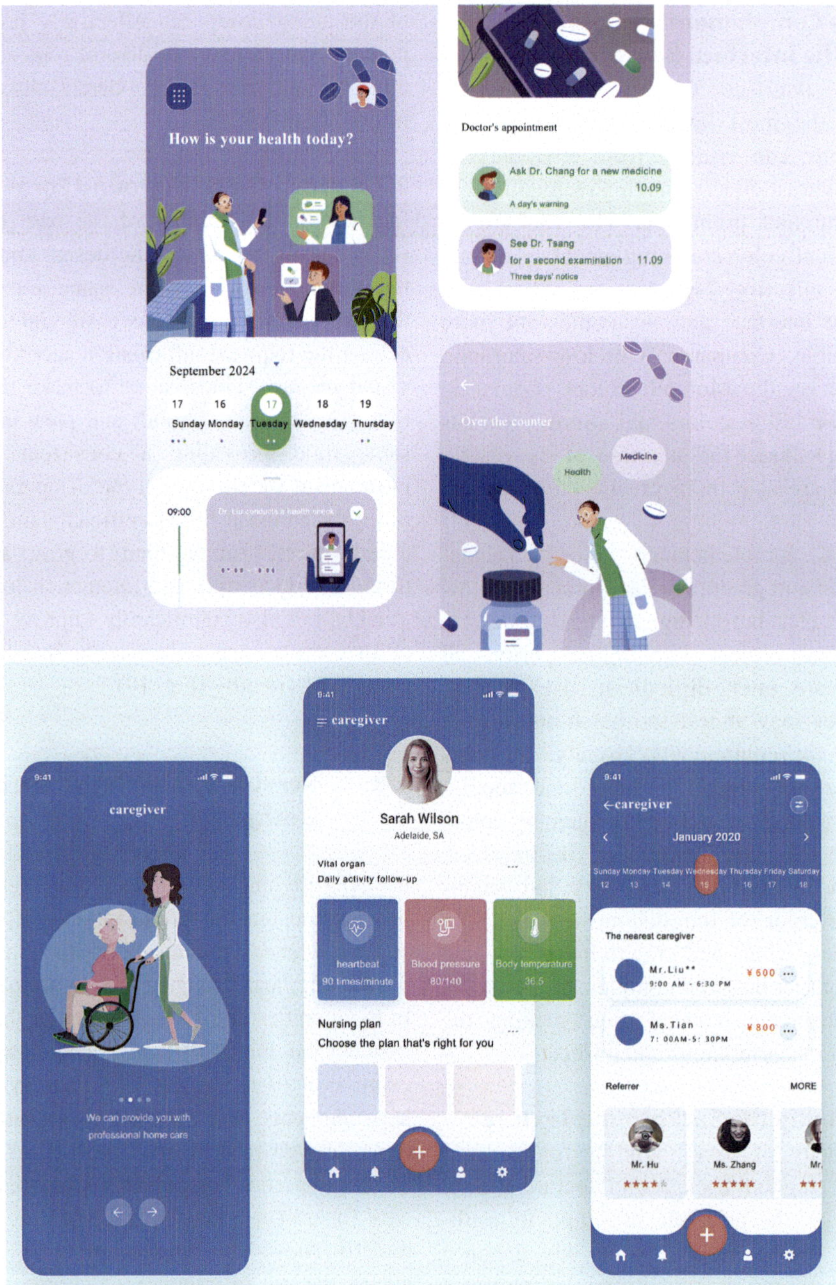


**Fig. 19** Case of an aging friendly app

### 2.8.1 Simple

The aging population generally faces a decline in visual contrast sensitivity, and may not be able to distinguish complex interface graphics, especially the ICONS on the screen of the mobile phone, whose display size is often limited to a very small range, such as the common 120 px × 20 px, or even smaller. Therefore, in the graphic design of the digital visual interface, it is necessary to use simple design elements as far as possible to avoid the graphics are not clear or even recognizable when displayed

**Fig. 20** Minimalist ICONS

in small sizes, and at the same time, it can also improve the graphic texture to a certain extent, so that the overall interface is more beautiful.

### 2.8.2 Direct

All design specifications explicitly state that icon design should be simple, friendly and highly recognizable. Usually the most direct way to adapt to aging is to increase the size of ICONS and buttons to achieve ease of operation and readability (Fig. 20). Another point is easy to ignore, as far as possible to use the design of skeuomorphism and universality, the icon expression as much as possible with the characteristics of the life times of the elderly, in line with their cognitive habits, the more concrete the icon, the more convenient for the elderly to use. While ICONS are powerful, it is equally important to use appropriate text to describe them (Fig. 21).

Considering the natural decline trend of logical reasoning and abstract thinking ability of the elderly group, icon design should pay more attention to fit the intuitive natural perception law, and strive to avoid the use of too obscure or highly abstract elements (Fig. 22). For example, in the weather application, the clouds directly symbolize the cloudy day and the sun represents the sunny day, which is intuitive and easy to understand. In the music playback software, musical notes are used as the core icon to directly convey music-related functions; When adjusting the volume, the amplified horn icon is paired with a clear increase or decrease sign to visually show the volume change.

### 2.8.3 Stable Type

As the aging population's ability to adapt to new things gradually slows down, they tend to take longer from initial exposure to proficiency in new technologies. This learning curve is more gradual than that of young people, and once mastered, they tend to maintain the status quo and do not readily accept big changes. Therefore, when designing digital visual interfaces for aging users, designers tend to adopt stable and easy-to-understand graphical representations. In the iterative process of system update, in order to maintain user habits, the changes of key elements such as ICONS, navigation bars, and controls are often controlled within a minimum range to ensure the consistency of the interface.

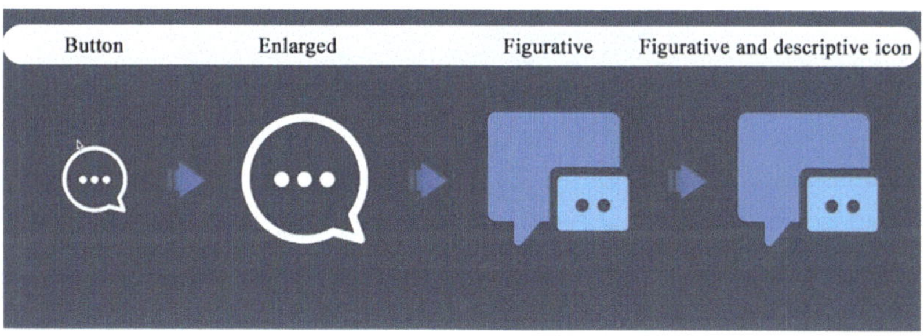


**Fig. 21** Concrete ICONS

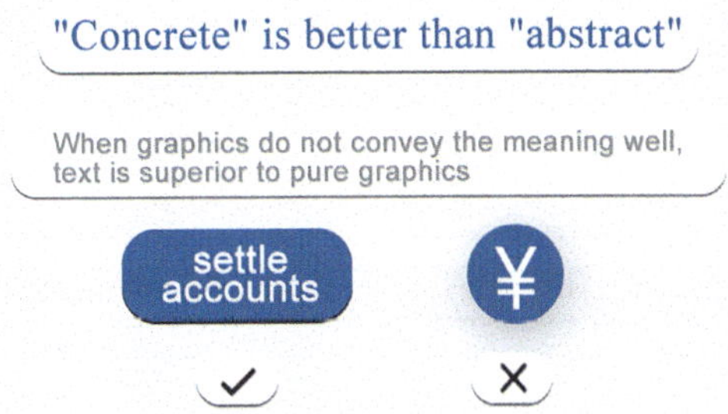


**Fig. 22** Avoid using ICONS that are not easily recognizable

## 3 Interface Layout and Navigation Design

In the digital visual interface design, the interface layout and navigation design should be closely focused on the user's intuitive operating experience, and strive to achieve the design goal of "clear, easy to learn, simple to operate, easy to use". The interface layout should be concise and clear, with reasonable distribution of function modules to avoid too many interfering elements. At the same time, the navigation design should have the characteristics of easy operation and significance, and establish a timely feedback mechanism to ensure that the digital interface is more in line with the cognitive level of elderly users, so as to achieve a barrier-free user experience (see Fig. 23).

### 3.1 Interface Layout Design

In the interface layout design, attention should be paid to the simplicity of the interface layout, and too many visual elements should be avoided as far as possible to ensure that users can quickly locate and understand the information (Fig. 24). The excessive use of Internet language, current buzzwords or local words may make it difficult for some users, especially the aging population or users from different cultural backgrounds, to understand and adapt. Such users are often unfamiliar with modern Internet language, and overly complex or cutting-edge words may lead to misunderstandings or omissions of information. In addition, the use of unusual traditional characters and other illegible symbols should be avoided to ensure that

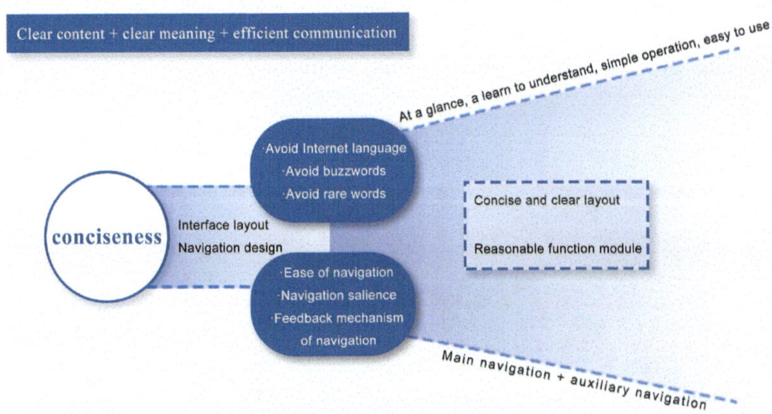


**Fig. 23** Interface layout and navigation design

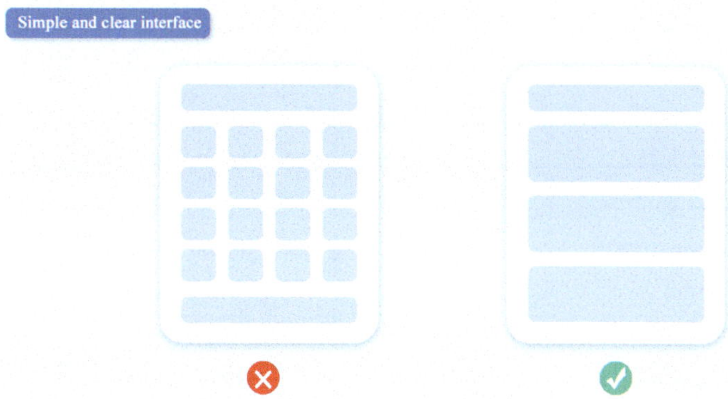


**Fig. 24** The interface layout should be simple and clear

all users can easily read and understand the text information in the interface.

### 3.1.1 Simplicity of Layout

For the aging population, when designing digital visual interface, the first principle is intuitive and easy to understand. Considering that this population is relatively weak in processing complex information, designers tend to build a single and clear interface environment, in which each element is simple and easy to identify, so as to enhance the intuitive operation. Too many functional options often confuse older users and make it difficult to make choices, while complex backgrounds can cause visual fatigue, distract attention and prevent the effective communication of key information. Therefore, in interfaces designed for the aging population, a streamlined strategy should be adopted to drastically reduce unnecessary feature options, remove those that are less frequently used or difficult to operate, and combine similar features to avoid redundancy. The interface layout focuses on preserving the core, most commonly used functions and creates a clean, uncluttered look with a simple and plain background (Fig. 25).

### 3.1.2 Language Expression of Information

As the earliest recording tool, text has a high accuracy. From the early text era of computers to today's graphical era, copy has always been the key to help us better understand information. On the limited screen of the mobile terminal, too long copy not only affects the visual effect, but also may lose the user's interest in reading, so choose concise and clear language as the copy. In the expression of information, the use of

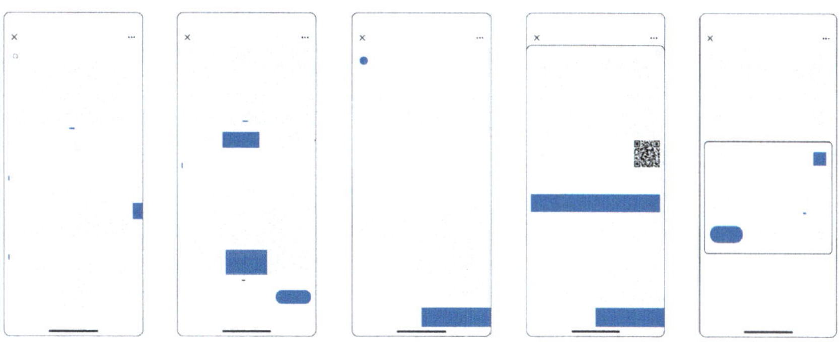

**Fig. 25** Layout diagram of a certain app

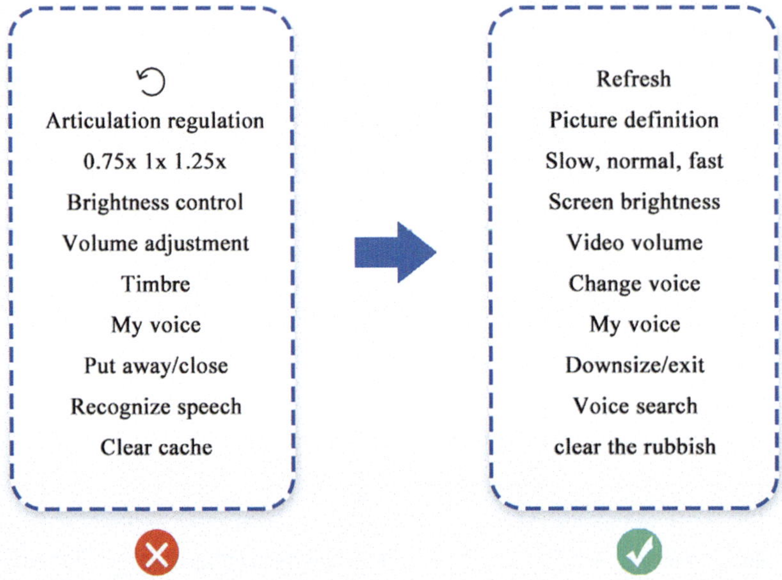


**Fig. 26** The information content is easy to understand

network language, popular words or rare traditional characters should be avoided to ensure the clear communication of information (Fig. 26).

## 3.2 Avoid Internet Language

As a new means of expression, network language has become an important part of digital life because of its simplicity, innovation, vitality and interest. However, the durability of Internet language remains to be tested. In digital interface design, designers should carefully avoid using network language. On the one hand, the aging group's ability to accept new things is relatively weak; On the other hand, due to the fluidity and uncertainty of network language, designers need to fully consider the language habits of the elderly and adopt clear and easy to understand forms of expression to reduce their cognition and understanding costs as much as possible.

## 3.3 Avoid Buzzwords

Language evolves with the continuous development of society, some words are given new meanings, while others are replaced by more novel expressions. For example, the term "Versailles", which originally referred to the luxurious palace of Versailles in France, is now used to describe people who seem to inadvertently brag about their privileged lives on social platforms by asking themselves questions before praising them. Due to the limited channels and ways for the elderly to obtain information, the speed of accepting new things is relatively slow, so they can not quickly grasp and understand the use of all popular words. In the digital interface design, if these popular words must be used, it is necessary to provide a clear explanation to avoid the elderly users in the operation of the process of misjudgment, thus affecting the correctness and smooth operation.

## 3.4 Avoid Unusual Traditional Characters

The evolution of Chinese characters shows a change trend from complexity to simplicity. However, in today's digital network, some rare and traditional characters with unique shapes and complex structures have unexpectedly

**Fig. 27** Common navigation forms

entered the public's vision. For the elderly group, considering the difference of their educational background, it is particularly important to use concise and easy to understand words or symbols. In addition, since many rare and traditional characters are not comprehensive, their display effect may vary on different devices, and this inconsistency will weaken the coherence and overall beauty of the interface, which will adversely affect the user experience. Therefore, in the process of design and communication, we should tend to choose more popular and compatible fonts, need to colloquial terms, and use popular and easy to understand copywriting to ensure the smooth transmission of information and the harmonious unity of the interface.

## 3.5 Navigation Design

Good navigation design can enable users to find the operation path clearly, and different navigation forms have their own characteristics in content display and interactive experience. With the continuous increase of product functional modules, it is difficult for a single navigation method to meet the needs of functional expansion. Therefore, the navigation design of most digital products usually adopts the mode of "main navigation + auxiliary navigation" to better support the complex functional structure (Fig. 27).

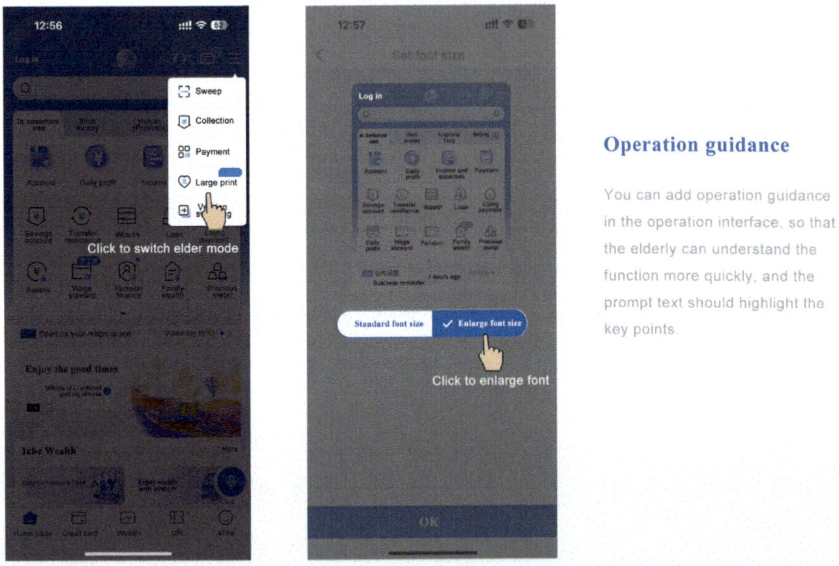


**Fig. 28** Operation guidance form

### 3.5.1 Ease of Navigation

In order to prevent older users from being unable to find a return path due to operational errors, or from affecting the operation process due to memory loss, an effective introductory tutorial should be provided to help users understand unfamiliar features (Fig. 28). Design with strict adherence to usability best practices, ensuring that the "back" button and "Home" navigation are easily accessible to meet the user's psychological expectations.

### 3.5.2 Navigation Salience

Navigation must have clear visual cues to suit their particular needs, and buttons and links should be designed to be as large as possible, easy to identify and easy to click. For aging users, the design needs to pay special attention to color contrast, ensure that there is enough visual difference between the navigation elements and the background, avoid using too similar or soft tones, especially those easily confused color combinations, such as light gray and white, which may make it difficult for users to accurately locate the functional area. By optimizing the navigation hierarchy to ensure that users find the information they need with the fewest clicks, they can significantly improve their experience. On the basis of expanding the visual prompt, it can also combine the dual prompt of the icon and the text to make the recognition of the function more intuitive. With concise and easy to understand language description and appropriate space white space, reduce the cognitive burden of users.

### 3.5.3 Feedback Mechanism for Navigation

Feedback and prompts play a crucial role in the user experience process. Information feedback includes not only the response to the operation command, but also the prompt of the system status. For older users, each click or touch should trigger a color change of the button so that they can clearly see that their actions are being recognized by the system. When the operation is successful, the system provides clear feedback through sound and animation effects to ensure that the user knows that the operation has been completed (Fig. 29). When false information or health content is involved, the navigation interface displays warning information in time. In addition, adding voice search and intelligent input functions, and providing sound feedback mechanisms, such as voice prompts when the operation is successful, can also significantly improve the operating experience of elderly users.

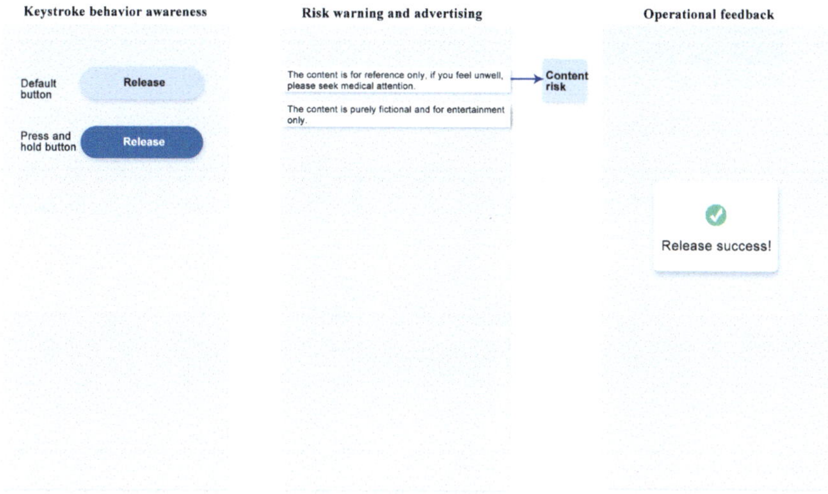


**Fig. 29** Form of operation feedback

# 4 Voice Interaction and Touch Feedback Design

In the interface design of the aging service system, voice interaction provides a more natural and convenient operation mode for the elderly users, avoiding cumbersome text input and complex menu navigation. Through clear and concise voice commands and instant voice feedback, the system can effectively improve the usability and response speed of interactions. At the same time, touch feedback design should focus on a simple and intuitive interface layout, combined with larger touch areas, significant visual cues (such as high-contrast color changes of buttons), and vibration and sound feedback to help older users better confirm actions. In general, the combined design of voice interaction and touch feedback can reduce the operational burden of elderly users and improve the experience of aging service systems (Fig. 30).

## 4.1 Voice Interaction Design

Voice recognition technology provides a convenient way to interact with the aging population, especially for those who are unfamiliar with traditional input methods or have degraded physical abilities. In reality, many elderly people due to the low level of culture, it is difficult to skillfully use the input method, and even do not understand pinyin, so most people choose handwriting input. However, handwritten input has serious problems with recognition accuracy. Coupled with the deterioration of physical function in the elderly, handwriting input is not an ideal solution, but a compromise option. With the help of artificial intelligence technology, voice input function can be provided for elderly users, so that they can more easily acquire knowledge through voice and image search, answer questions, and keep up with the pace of The Times.

### 4.1.1 Accuracy of Speech Recognition

Older adults may encounter several unique challenges when using speech recognition. First, with age, older people may no longer pronounce as clearly as they did when they were younger, and the speed of speech may also become slower or uneven, making it difficult for the system to accurately understand their commands. In addition, dialects or accents of different regions will also increase the difficulty of speech recognition, and the system's misrecognition rate may increase significantly. In order to solve these problems, speech recognition technology needs to be continuously optimized to improve the

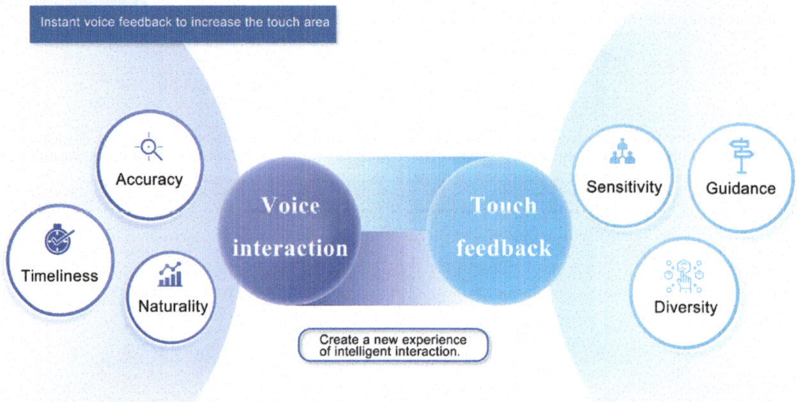


**Fig. 30** Design of voice interaction and touch feedback

adaptability to the speech features of the elderly. For example, systems should be designed with stronger fault tolerance in mind, as well as personalized training models to help the system better recognize speech patterns of older users (Fig. 31).

### 4.1.2 Timeliness of Voice Feedback

For the aging population, speech recognition technology must not only have high accuracy, but also the timeliness of voice feedback is crucial. Older users, when using voice input, often rely on real-time system feedback to confirm that their commands are being performed correctly. However, in the process of voice interaction, the elderly often have high expectations and dependence on the system response speed, if the voice feedback is too slow, it is easy to make them feel anxious, and even think that the system does not correctly recognize their instructions. In addition, too long a feedback delay may cause them to issue the same instructions repeatedly, further increasing the risk of misoperation. In order to improve the experience of older users, voice recognition systems should minimize feedback delays and ensure immediate responses. Even during complex task processing, the system can let the user know that the instruction has been received through intermediate feedback, such as confirmation of what has been heard.

### 4.1.3 Naturalness of Voice Interaction

The elderly are used to the natural way of conversation in daily life, and if the interaction of the voice recognition system seems stiff or too mechanical, it may cause their experience to become not smooth, and even make them reject the technology. Natural voice interaction not only requires the system to be able to accurately recognize instructions, but also needs to imitate human language habits during the dialogue process and have flexible dialogue capabilities. For example, older people may use less standard language or express lengthy instructions, and the system must be able to understand these inputs, rather than requiring the user to follow a fixed format. In addition, voice interaction systems should have context understanding and be able to respond coherently based on the context of the conversation, without requiring the user to repeat or re-enter instructions frequently. At the same time, the system should minimize unnecessary technical terminology, so that the interaction process is more close to the language style of daily life.

## 4.2 Touch Feedback Design

As the elderly age, their tactile sensitivity and hand-eye coordination gradually decline, which makes them prone to misoperation or slow

**Fig. 31** Huawei's smart voice visual image design

reactions when operating touch screen devices. Therefore, a good touch feedback design can effectively help elderly users improve the accuracy and confidence of operation. The design of touch feedback should avoid overly complicated multi-point gesture operation. Intuitive single-touch and sliding operation are more suitable for the elderly, and complex gestures may increase their cognitive burden and affect the overall user experience (see Fig. 32).

### 4.2.1 Sensitivity of Touch Feedback

When designing the feedback effect, first of all, the size of the feedback information should be increased, and the prompt time should be extended and the feedback intensity should be enhanced. For example, the regular prompt pop-up display time is 2 s, and in the aging interface, it should be extended to 5 s. In addition, the click operation that lacks vibration feedback under normal circumstances should be included in the age-appropriate design to help elderly users better perceive the completion of the operation.

In the design of the touch control area, the operation habits and physiological restrictions of elderly users should be fully considered. Larger touch buttons and wider clicking areas can reduce misoperation, so that the elderly do not need too fine finger control (see Fig. 33). At the same time, the spacing between the buttons should be appropriately increased to avoid miscontact caused by too dense buttons. Each touch operation should be accompanied by clear

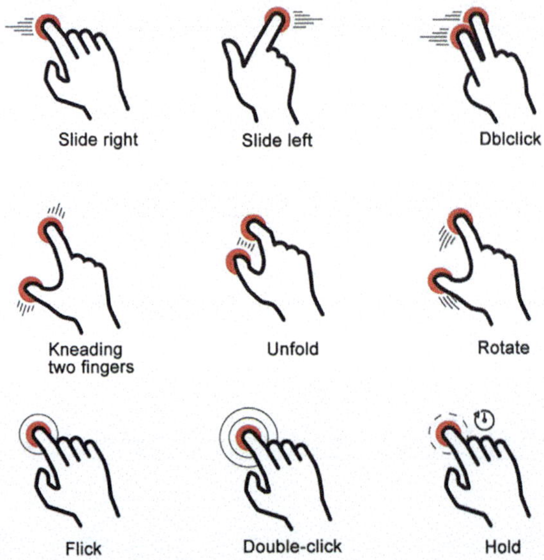


**Fig. 32** Commonly used interaction gestures

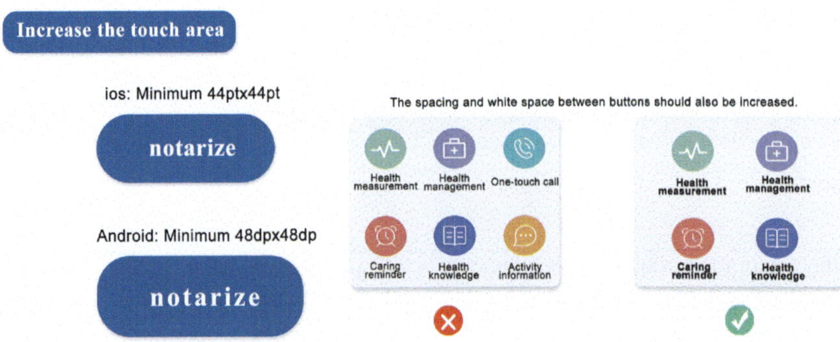


**Fig. 33** Increase the touch area of the button

feedback, such as button color changes, slight amplification or vibration prompts, to ensure that the user knows that the system has successfully received the command.

### 4.2.2 Guidance of Touch Feedback

For the aging population, interface design not only needs to provide clear feedback, but also needs to be guided through reasonable. First of all, the guidance of the feedback should be reflected in the step-by-step prompts of the operation steps. For example, in a multi-step operation, the system can gradually guide the user to complete the task through touch feedback. Secondly, the guidance of touch feedback should also be reflected in the correction of wrong operation. When elderly users have wrong touch or wrong operation, the interface should give clear prompts to guide them back to the correct operation path through touch feedback. For example, provide a prominent error box, vibration feedback or voice reminder, and clearly tell how to correct the action.

### 4.2.3 Diversity of Touch Feedback

Touch feedback includes not only visual changes, but also enhanced perception through sound and vibration. For example, buttons can change color, zoom in, or lighten when clicked, helping users clearly see that an action has taken effect. At the same time, sound feedback is an effective guidance tool. Whenever the user completes an action, such as opening an app or hitting the back key, the system should provide a clear audible notification that the action was successful. Vibration feedback can also be supplemented, when the user long presses or double clicks the screen, the device can confirm that the action has been recognized through a slight vibration, providing more explicit operational feedback.

# 6 Optimization of Interactive Experience

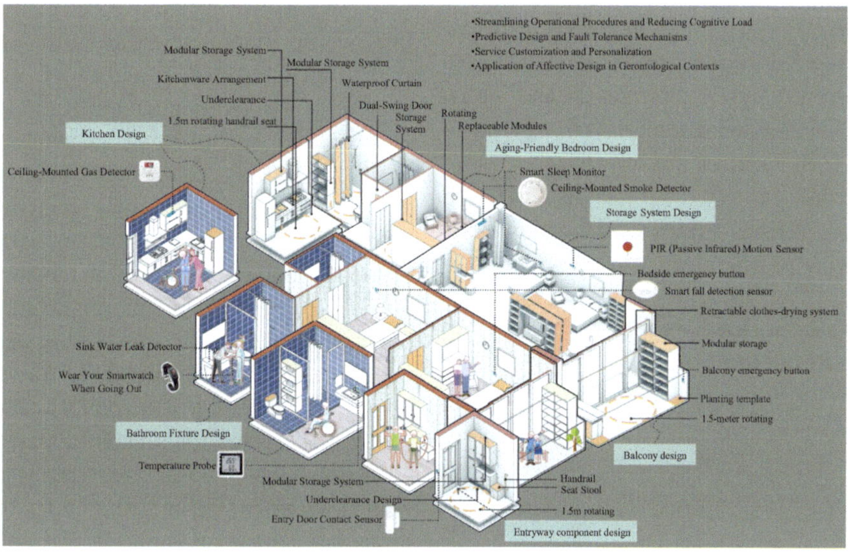


## 1 Streamlining Operational Procedures and Reducing Cognitive Load

### 1.1 Theoretical Foundations of Cognitive Load

#### 1.1.1 Definition and Taxonomy of Cognitive Load

When users acquire knowledge or solve problems, cognitive processing consumes cognitive resources. Cognitive load (Cognitive Workload) refers to the multidimensional burden imposed on users' cognitive systems during such cognitive processing. Within the causal dimensions of cognitive load, task characteristics, learner characteristics, and their interactions collectively determine the sources of cognitive load. When task demands remain within the cognitive resource capacity of elderly users, they can complete tasks effectively. As task complexity increases and exceeds cognitive resource thresholds, overload occurs, leading to a significant decline in task performance.


W. Tang, *Aging-friendly Service System Design*, https://doi.org/10.1007/978-981-95-2054-1_6

Cognitive load is typically categorized into three types: intrinsic cognitive load, extraneous cognitive load, and germane cognitive load. Extraneous cognitive load arises from the complexity of information presentation formats and task rules. When information materials are designed in a complex or disorganized manner, users encounter obstacles in completing tasks smoothly. Intrinsic cognitive load is closely related to the complexity and quantity of information materials, as well as individual knowledge and capabilities. When the complexity or volume of processed information exceeds users' cognitive capacity, intrinsic cognitive load increases substantially. Germane cognitive load is associated with the allocation of mental effort during learning. Users' long-term memory permanently stores information and skills, organized in units of schemas. When germane cognitive load decreases, users allocate more effort to processing irrelevant factors, leading to increased extraneous cognitive load and reduced task efficiency.

### 1.1.2 Strategies and Methods for Reducing Cognitive Load

Extraneous cognitive load stems from the interaction between individual capabilities and specific tasks. In interface design, optimizing information organization, simplifying interaction processes, and enhancing the clarity of information presentation aim to reduce extraneous cognitive load, thereby enabling users to allocate cognitive resources more effectively to task execution.

Methods to reduce intrinsic cognitive load include summarizing information through graphical and modular approaches to improve users' cognitive efficiency, minimizing interactions between elements to lower recognition difficulty, and simplifying information structures and hierarchies to facilitate rapid retrieval and access.

In designing for germane cognitive load, universal schemas should first be employed to allow users to quickly match new information with existing schemas, enabling rapid extraction and application of information. Second, efforts should focus on helping users construct schemas to accelerate their transition from novices to experts. Therefore, to enhance cognitive performance, designs should prioritize reducing users' intrinsic and extraneous cognitive loads while increasing germane cognitive load.

## 1.2 Core Principles and Methods of Design Simplification

### 1.2.1 Information Simplification and Enhancement of Visual Clarity

Information simplification and enhanced visual clarity are core strategies for optimizing user interfaces. Information simplification involves refining and summarizing content to ensure users can quickly comprehend and effectively utilize information. Specific methods include graphical representations, such as using charts, diagrams, and flowcharts to replace lengthy textual explanations. This approach reduces reliance on textual information while improving readability and memorability (as shown in Fig. 1). Additionally, the application of visual hierarchy (e.g., headings, subheadings, lists) and contrast (e.g., color, font size, weight) highlights critical information, enabling users to rapidly locate key content (as shown in Fig. 2).

### 1.2.2 Principles and Practices of Operational Process Optimization

Operational process optimization focuses on enhancing task completion efficiency and reducing operational complexity. Key strategies include minimizing operational steps and interactions through step consolidation or shortcut provision, which significantly streamlines workflows and improves task execution fluency. Maintaining interface consistency and standardization reduces learning costs, enabling users to rapidly adapt to new operational environments. Allowing undo/redo operations enhances control

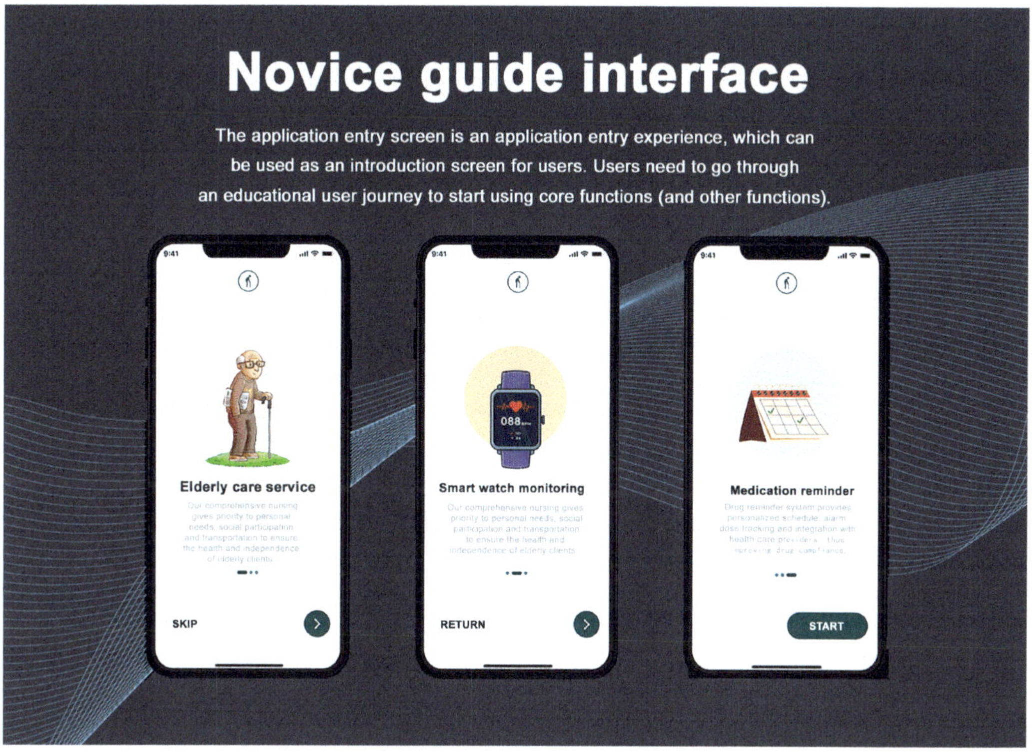


**Fig. 1** Graphical operation flow

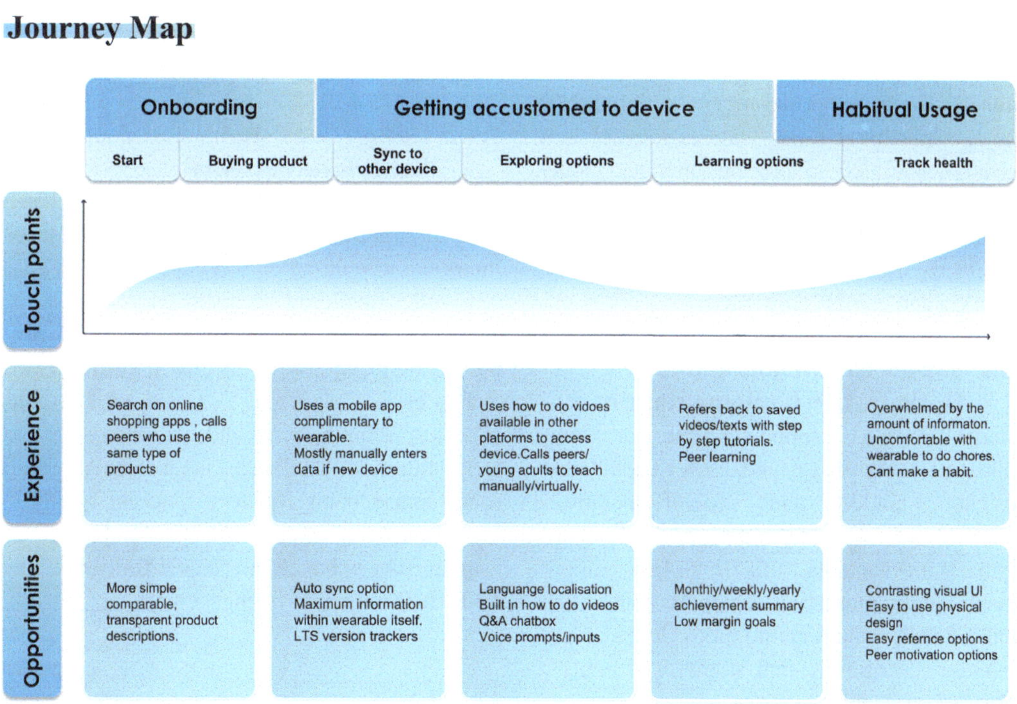


**Fig. 2** User journey map

perception while mitigating error- related burdens. Visualized workflows (e.g., flowcharts or step indicators) provide clear task overviews, facilitating operational continuity. The integrated application of these optimization s substantially enhances interface usability and operational efficiency.

## 1.3 User Task Analysis and Interface Optimization

### 1.3.1 Task Decomposition and Operation Prioritization

In gerontological service system design, task decomposition proves critical for optimizing elderly users' operational experience. Given older adults' limited cognitive and operational capacities, breaking complex tasks into sequential steps reduces cognitive load and enhances task fluency. Decomposition must align with elderly users' behavioral patterns, prioritizing analysis of high-frequency functions to ensure essential tasks receive prominence and simplification.

Operation prioritization should adjust based on task criticality, complexity, and frequency. High—priority operations (e.g., daily service functions) require prominent positioning for immediate access, while low-priority features may be strategically concealed to prevent interface clutter. Through systematic task decomposition and prioritization management, interfaces better accommodate elderly users' operational habits, minimizing confusion and barriers during use.

### 1.3.2 Support Strategies and Task Optimization Implementation

Elderly—oriented support strategies must emphasize usability and user-friendliness. Designers should provide explicit operational guidance, salient visual cues, and responsive feedback mechanisms to facilitate task comprehension and completion. Support frameworks should incorporate unambiguous icons, plain-language descriptions, and stepwise prompts to reduce anxiety and error rates.

Task optimization implementations must adapt to elderly needs through interface simplification, redundant interaction elimination, and visual distraction reduction. Particularly during complex task execution, minimize interface navigation frequency while ensuring operational path clarity. Adaptive interface design further enhances optimization by automatically adjusting workflows and elements based on user behavior patterns.

## 1.4 Interface Design Impacts on Cognitive Load

### 1.4.1 Impact of Interface Layout on Cognitive Load

Aging-related declines in perceptual acuity and cognitive processing make complex layouts particularly detrimental to elderly users, increasing cognitive load and impeding information processing. Clear, minimalist layouts reduce information search/processing time, effectively lowering cognitive demands. Critical information and functions must occupy visually dominant positions, guided by explicit visual hierarchies to direct attention and accelerate content retrieval (as shown in Fig. 3).

### 1.4.2 Design Consistency and User Behavioral Patterns

Design consistency must align with elderly users' experiential operational preferences. Interfaces should adhere to cognitively compatible rules, avoiding frequent view transitions and multi—step complexities while prioritizing accessibility of frequently used functions. Consistency permeates typography, color schemes, iconography, and interaction behaviors. Intuitive feedback mechanisms ensure users comprehend system responses, bolstering confidence and experience. Customizable settings further enhance interface affinity and usability, fostering elderly engagement.

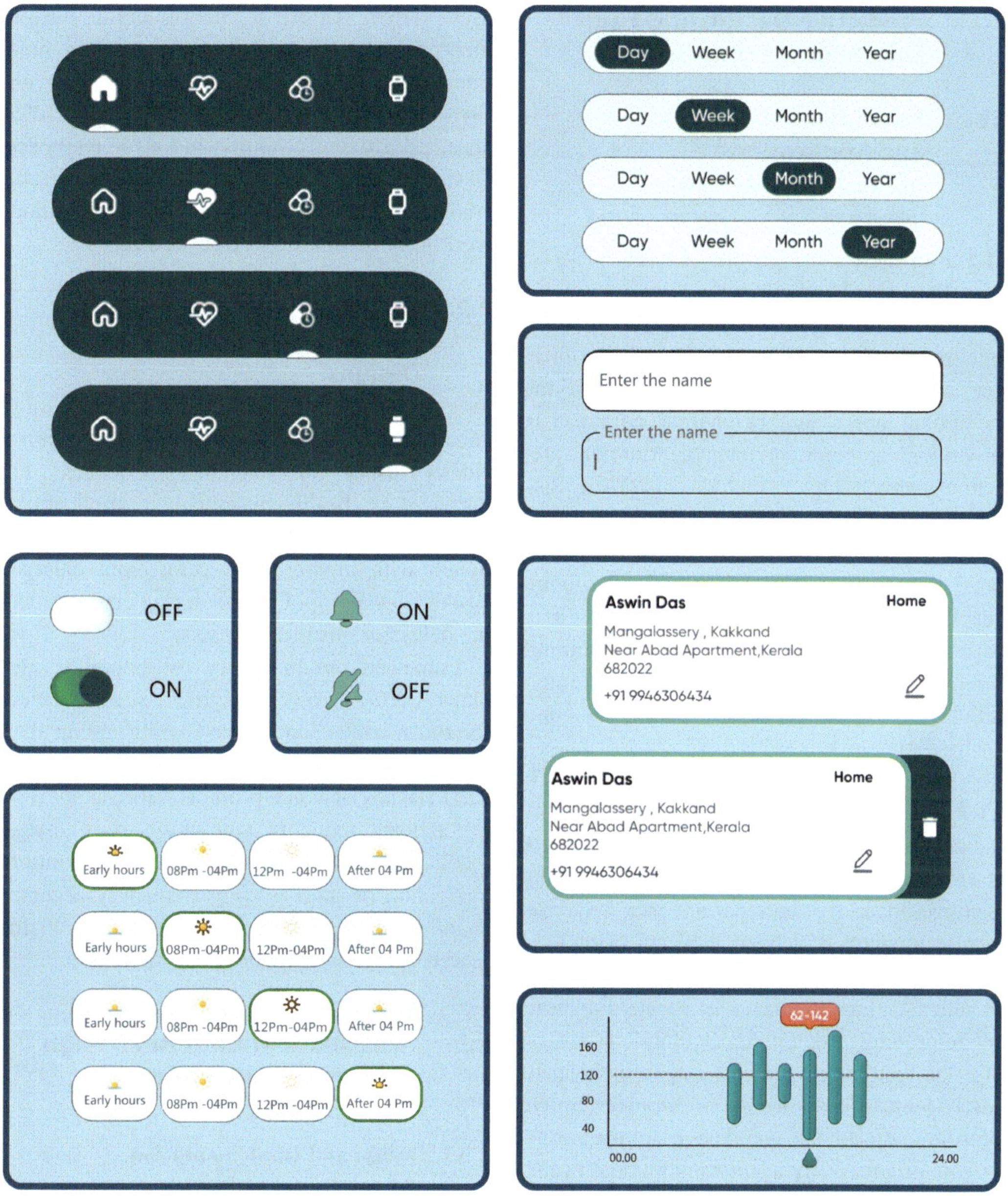


**Fig. 3** Custom settings interface

## 2 Predictive Design and Fault Tolerance Mechanisms

### 2.1 Principles and Applications of Predictive Design

#### 2.1.1 Fundamental Methods of User Behavior Prediction

Predictive design, grounded in the analysis and anticipation of user behavior, aims to optimize user experience by preemptively identifying operational needs and behavioral patterns. In gerontological service systems, common prediction methodologies include historical data analytics, machine learning algorithms, and behavioral pattern modeling. By analyzing older users' operational habits, preferences, and common behavioral pathways, these methods enable systems to anticipate subsequent actions, thereby optimizing response speeds and interface layouts while reducing operational complexity during task execution.

#### 2.1.2 Benefits of Predictive Design

The primary advantage of predictive design lies in enhancing user experience and mitigating cognitive load. By anticipating user behavior, systems proactively deliver relevant information or operational suggestions, eliminating the need for elderly users to manually locate functions and improving task efficiency. Implementation relies on continuous analysis and refinement of elderly users' operational data. Through iterative algorithm adjustments and design optimizations, systems progressively adapt to individual needs, thereby enhancing intelligent and personalized service capabilities.

### 2.2 Core Concepts of Fault Tolerance Mechanism Design

#### 2.2.1 Definition and Strategies of Fault Tolerance Mechanisms

Fault tolerance mechanisms refer to design strategies that allow systems to tolerate and rectify user errors. Effective fault-tolerant designs incorporate undo operations, auto-save functionality, and multi-step confirmation protocols. Error prevention remains central to fault tolerance, prioritizing usability and error resilience. Designers must strategically plan functional zoning, intuitive indicators, and streamlined workflows to minimize error occurrence.

#### 2.2.2 Practical Applications of Common Fault Tolerance Designs

In gerontological service systems, fault tolerance implementations significantly improve elderly users' operational experiences. For instance, in health management applications, emergency call buttons are prominently positioned with high-contrast colors and enlarged sizes to ensure rapid access during crises, reducing delays or misoperation risks.

Enhanced multisensory stimulation also plays a critical role. In digital assistant devices for older adults, operational confirmation integrates visual and auditory feedback—green LED flashes or voice prompts validate successful actions, reducing ambiguity from unclear feedback. Similarly, smart TV remotes simplify navigation by auto- sorting frequently accessed channels at the interface forefront, minimizing search efforts and repetitive interactions.

### 2.3 Feedback Mechanism Design and Error Correction

#### 2.3.1 Design and Implementation of Real-Time Feedback

Real-time feedback mechanisms are critical for ensuring elderly users comprehend system response s and execute correct operations. Feedback should integrate multimodal stimuli (visual, auditory, or tactile) to amplify operational outcome perception. For example, successful actions may trigger color transitions, auditory alerts, or haptic vibrations to confirm validity.

### 2.3.2 Impact of Feedback Mechanisms on Error Correction

When errors occur, systems must immediately provide explicit notifications detailing error causes and rectification steps. For instance, erroneous data input may trigger pop-up alerts or highlighted text specifying the issue and offering concise resolution guidance, preventing user confusion (as shown in Fig. 4).

## 3 Customization and Personalized Services

### 3.1 Theoretical Foundations of Customization and Personalization

Customization and personalized services constitute a user-centered service model for elderly users, analyzing their interests and habits to

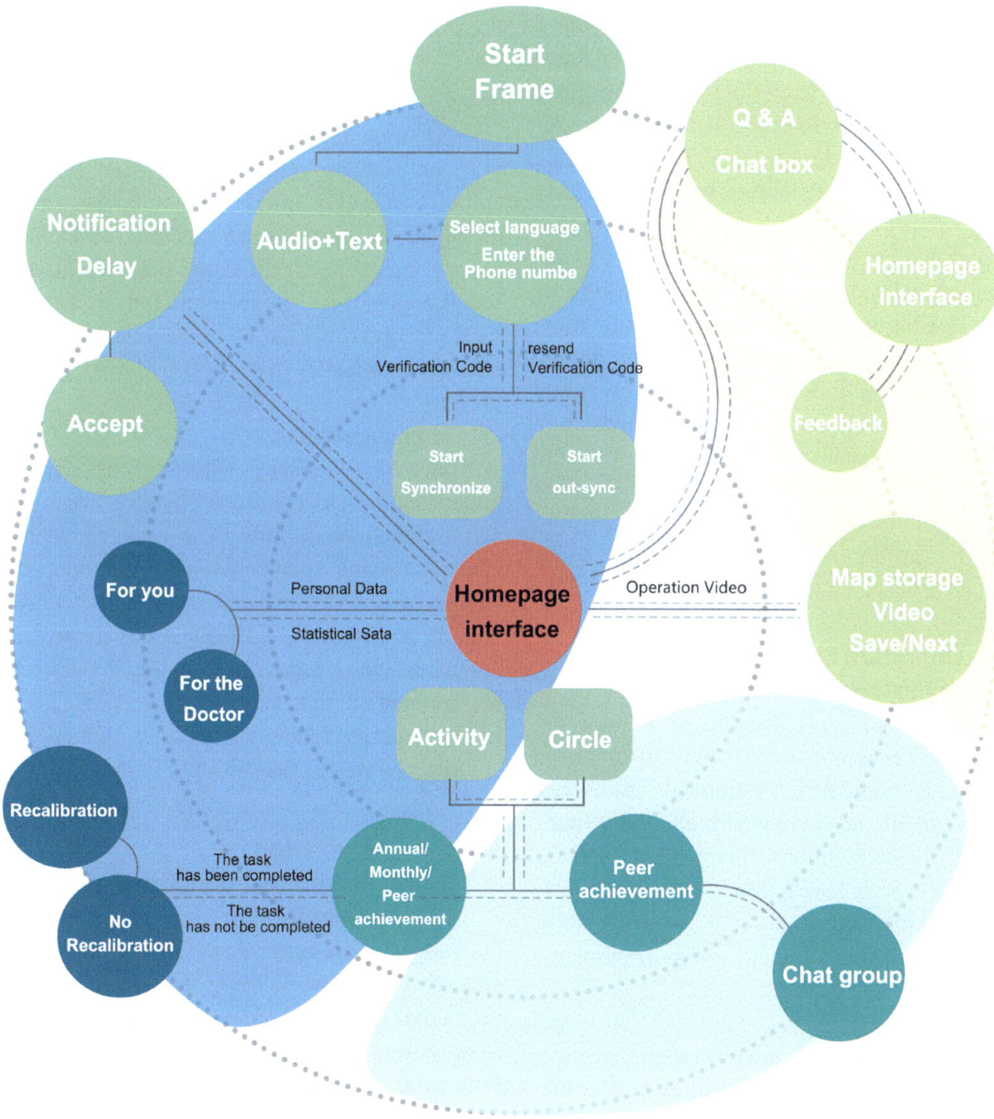


**Fig. 4** Interface design of feedback mechanism

proactively deliver exclusive information and services through user-defined settings, system recommendations, and push notifications. Core characteristics include relevance, proactivity, and intelligence. These services actively analyze elderly users' needs, delivering timely information. As user data accumulates and updates, systems continuously refine their recognition accuracy through user modeling, enabling increasingly precise service delivery.

### 3.1.1 Core Principles and Theories of Customized Design

In gerontological service systems, customized design revolves around providing personalized solutions to meet diverse elderly needs. Rooted in human-computer interaction and user experience design theories, it emphasizes adapting interfaces and functionalities based on elderly users' operational habits, cognitive abilities, and physical conditions. Customization extends beyond visual and interface adjustments to functional flexibility. For example, in health management platforms, elderly users can define personalized monitoring metrics (e.g., heart rate thresholds, blood pressure ranges, or medication reminder frequencies) tailored to their health conditions. Such personalized configurations reduce operational complexity, enhance engagement in comfortable environments, and foster trust and reliance on the system.

### 3.1.2 Definition and Evolution of Personalized Services

Personalized services dynamically adjust content using behavioral data, usage history, and preferences to deliver tailored experiences. Elderly-specific needs in health and daily living drive intelligent solutions. Smart home systems, for instance, learn user routines to autonomously adjust lighting, temperature, or security alerts, minimizing manual inputs while optimizing comfort.

Advancements in big data and AI propel personalized services toward deeper refinement. Medical devices and wearables now offer real-time health monitoring coupled with customized advice (e.g., exercise plans, dietary adjustments). These innovations enhance system intelligence, convenience, and user-centricity, improving elderly independence and quality of life.

## 3.2 User Needs Analysis and System Design

### 3.2.1 Theoretical Methods for User Needs Analysis

User needs analysis is fundamental to ensuring systems effectively address elderly users' diverse requirements. Common theoretical methods include user research, behavioral analysis, and needs prioritization. User research employs surveys, interviews, or focus groups to collect qualitative and quantitative data on elderly users' needs in daily life, health, and social engagement. Behavioral analysis observe s elderly users' operational habits and cognitive patterns to identify pain points during system use. By prioritizing needs, designers optimize critical functionalities.

Theoretically, user needs analysis requires understanding elderly users' physical, cognitive, and psychological traits. Age—related decline s in processing speed, memory retention, and sensory responses necessitate special consideration. Applying hierarchical needs theory and contextual design theory, designers dissect core requirements across multiple dimensions, translating them into functional modules aligned with elderly preferences (as shown in Fig. 5).

### 3.2.2 Customization and Personalization Strategies in System Design

Guided by user needs analysis, system design strategies aim to deliver interaction experiences tailored to elderly users' unique traits. Customization strategies enable interface and functional adjustments based on preferences,

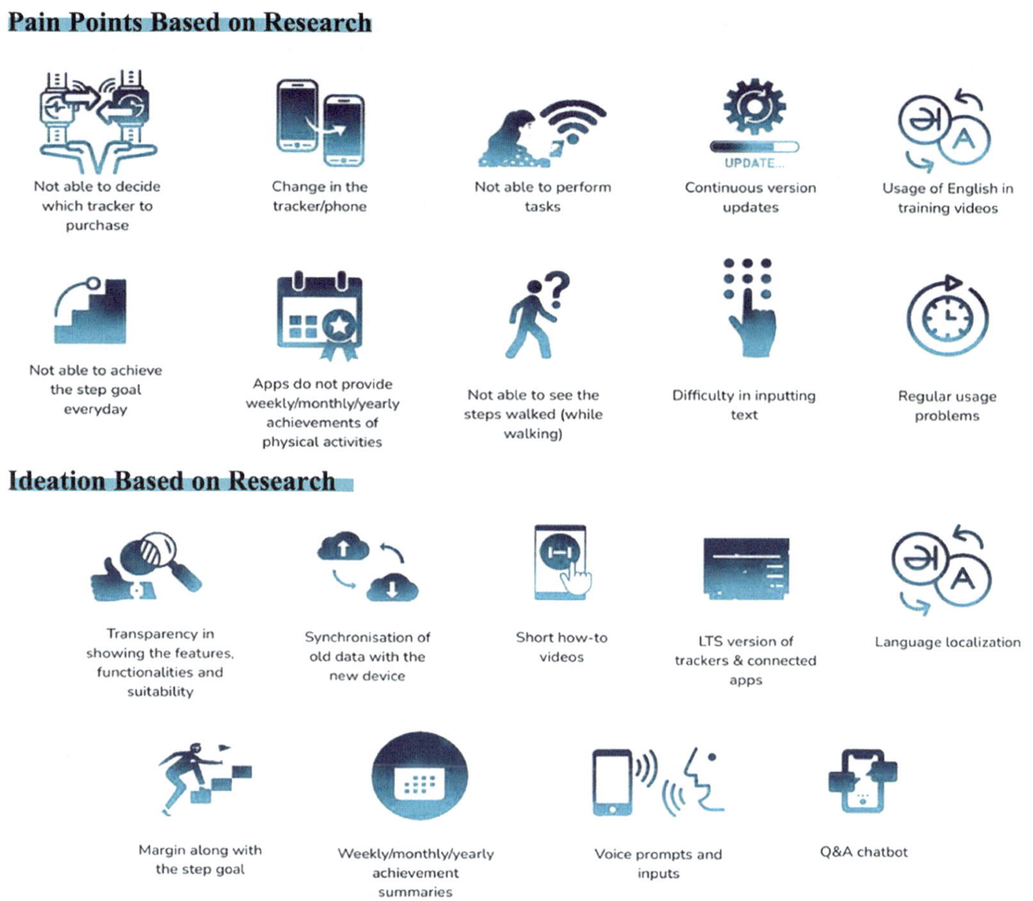


**Fig. 5** Personalised strategy for the use of products by elderly users

habits, and capabilities. For example, modular designs allow font enlargement and step reduction for simplified information access, ensuring seamless usability.

Personalization strategies leverage big data and AI to dynamically adapt service content and interactions. Health monitoring systems, for instance, auto- adjust measurement frequencies and generate personalized recommendations based on health data trends. Intelligent recommendation algorithms push relevant content aligned with user interests, boosting engagement and satisfaction.

## 3.3 Practical Applications of Customized and Personalized Services

### 3.3.1 Personalized Interface Design in Elderly Financial Management Systems

Elderly financial management systems feature highly customizable interfaces to simplify complex financial operations. Users set personalized reminders for pensions, savings balances, or budget controls (as shown in Fig. 6). The system generates tailored budgeting and savings advice

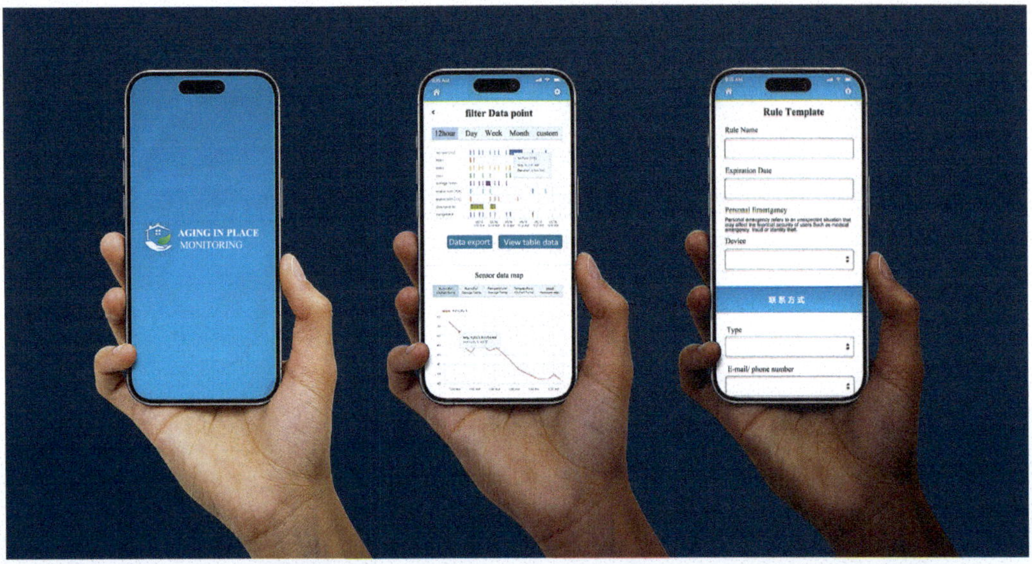


**Fig. 6** Personalized interface design of financial management system for the elderly

based on financial status. Interfaces employ large fonts, simplified menus, and high-contrast charts for quick financial tracking, reducing cognitive load in decision-making and ensuring economic security (as shown in Fig. 7).

### 3.3.2 Smart Home System Customization

Smart home control panels offer elderly users convenience in household management through customizable automation. Users configure appliances based on preferences—e.g., optimizing lighting to reduce nighttime glare or activating security monitoring during specific hours (as shown in Fig. 8). Intelligent learning features adjust home environments autonomously, such as preheating rooms at habitual wake-up times. These automated settings reduce operational burdens while enhancing safety and convenience. User-friendly interfaces with clear layouts further ensure accessibility, aligning with elderly cognitive traits.

### 3.3.3 Personalized Health Monitoring Interfaces

In Elderly Health Management Systems, customized interfaces employ personalized design to address the health monitoring needs of elderly users. For instance, the system enables users to select monitored health indicators such as blood pressure, blood glucose, or body weight. Users may configure personalized reminders and data recording intervals based on individual health conditions. The system dynamically adjusts monitoring frequencies and alert settings through continuous analysis of daily biometric data, with prioritized visual highlighting of health anomalies when critical thresholds are detected (see Fig. 9).

Within Personalized Elderly Care Services, the paradigm of elderly care is evolving toward hyper-personalization through demand-driven system design. This model prioritizes the granular understanding of elderly users' multidimensional needs, leveraging advanced analytics of physiological, psychological, and behavioral datasets to architect service architectures that precisely align with their lifestyle patterns and health profiles. Compared to traditional standardized service models, personalized elderly care frameworks demonstrably enhance system dependency and user satisfaction among older adults, thereby elevating overall quality of life.

Through Customization-Driven Design, elderly care systems deliver highly individualized health management and living support

**Fig. 7** Interface design of financial management system for the elderly

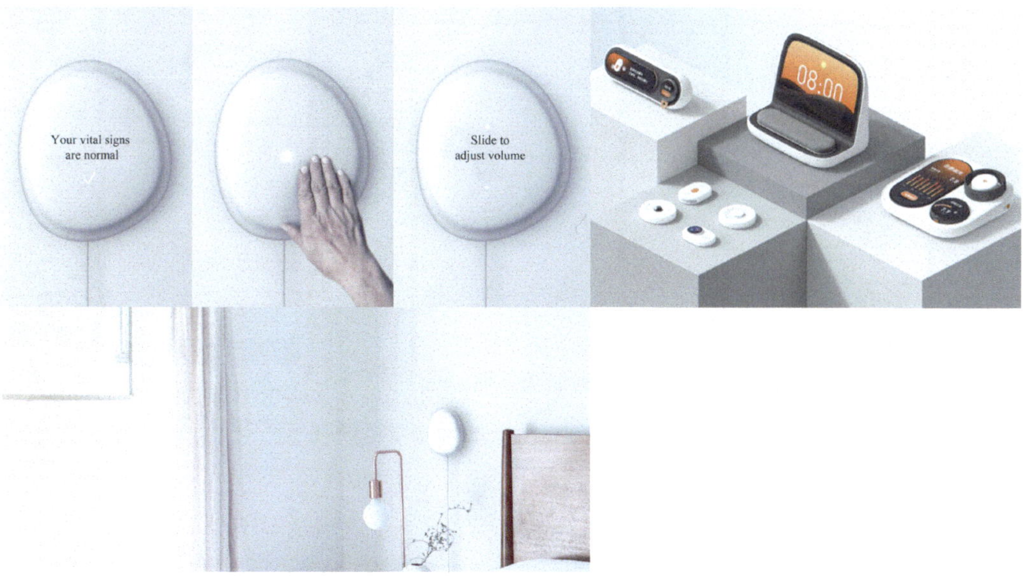


**Fig. 8** Personalized interface design of smart home system

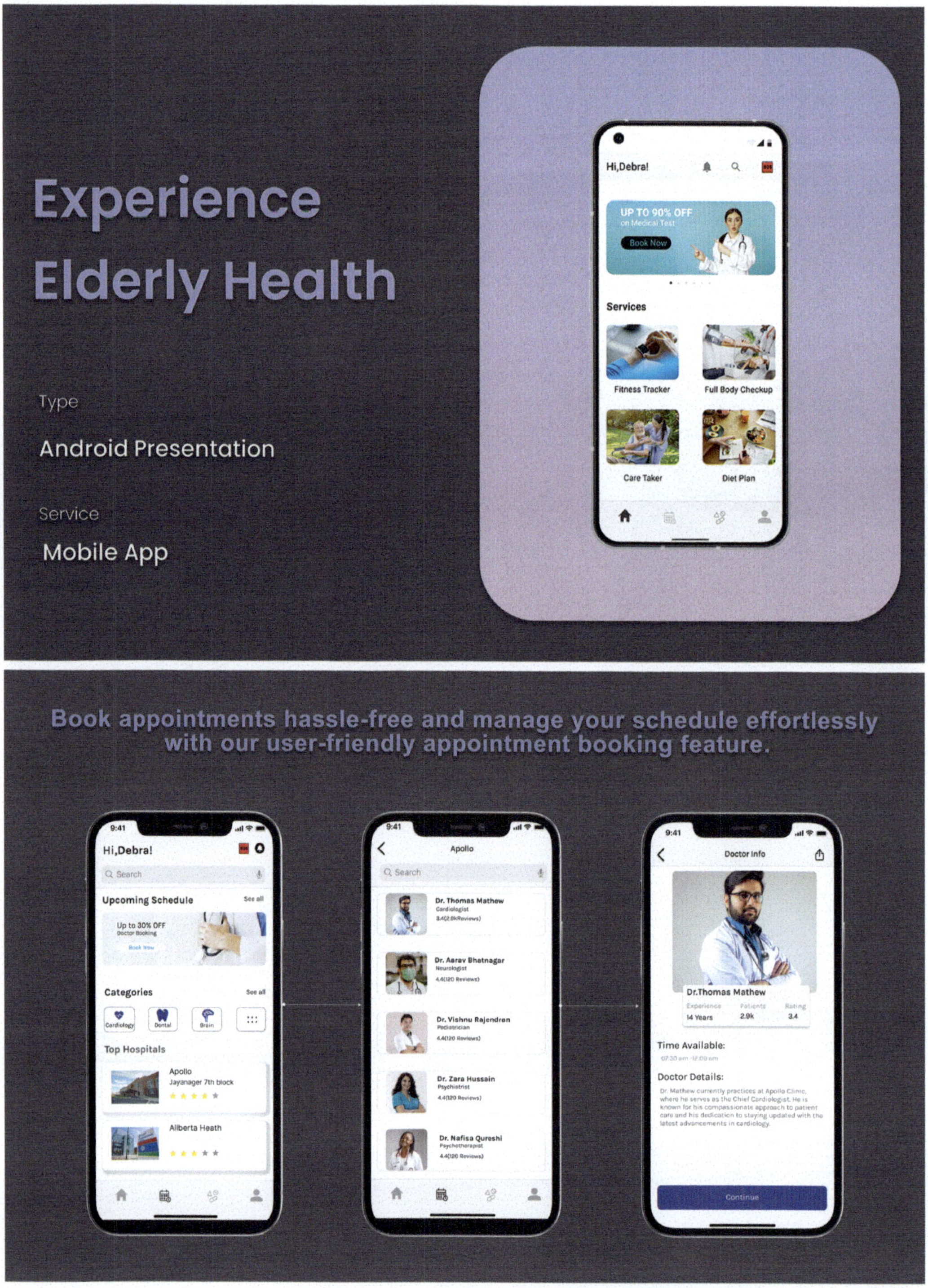


**Fig. 9** Personalized interface design of elderly health management system

tailored to specific user requirements. In the health management domain, personalized services utilize geriatric health data to formulate bespoke monitoring protocol s and precision interventions, strengthening seniors' capacity for autonomous health governance. For home -based care scenarios, smart home systems employ adaptive customization to autonomously optimize living environments, ensuring seamless environmental adaptation and proactive safety reinforcement. This user-centric design philosophy guarantees superior system adaptability and operational accessibility, particularly crucial for elderly populations with higher cognitive loads, effectively reducing operational complexity and psychological strain.

Beyond Health and Daily Support, personalized services extend into social engagement and mental wellness domains. Algorithmically optimized virtual companionship platforms and context-aware social interfaces maintain elderly users' societal connectivity through intelligent content curation and community matching, thereby enhancing psychological well -being and life fulfillment.

At Its Core, the personalized elderly care model transcends reactive physiological support, constructing holistic care ecosystems through multimodal data integration and predictive analytics to accommodate dynamic, complex aging needs. By positioning elderly users' active participation as the central design imperative, this paradigm drives the intelligent, human-centered evolution of future elderly care systems.

# 4 Application of Affective Design in Gerontological Services

## 4.1 Definition and Theoretical Foundations of Affective Design

### 4.1.1 Definition of Affective Design

Affective design refers to the creation of emotionally resonant products and services through the interaction of design elements with users' emotions. This design philosophy prioritizes users' emotional needs, leveraging color, form, material, and interaction modalities to elicit positive emotional responses. Theoretically, affective design is rooted in psychological and user experience research, emphasizing how emotional resonance enhances user satisfaction and product value.

Affective design draws on multidisciplinary theories, particularly cognitive psychology, user experience studies, and human-computer interaction. Renowned design scholar Don Norman's Three Levels of Emotional Design theory posits that products should engage users emotionally across three dimensions: visceral (immediate sensory perception), behavioral (operational experience), and reflective (deeper cognitive engagement). By addressing these levels, affective design improves usability while fostering enduring product appeal and market competitiveness. This approach is increasingly applied across digital products, service systems, and physical goods, creating more human-centric experiences.

### 4.1.2 Impact of Affective Design on User Experience

Affective design enhances user experience by elevating emotional satisfaction and overall engagement. Thoughtfully crafted interfaces, interactions, and visual elements evoke positive emotions such as pleasure, comfort, and belonging. For example, intuitive interfaces and responsive designs mitigate user frustration, while warm color palettes and personalized feedback boost satisfaction and loyalty.

Affective design also profoundly influences long-term user behavior. Positive emotional experiences increase product satisfaction and foster habitual use. Studies demonstrate that users receiving and affirmative feedback during interactions are more likely to sustain usage and recommend products to others. This emotional bond strengthens brand loyalty and cultivates market goodwill. Thus, affective design not only elevates immediate satisfaction but also enhances market performance and long-term loyalty through emotional resonance and virtuous user experience cycles.

## 4.2 Affective Design Strategies for Aging Populations

### 4.2.1 Gerontological Affective Design Principles

Affective design principles for elderly users address their unique physiological, psychological, and social needs. Key principles include Simplifying interfaces for intuitive usability to reduce cognitive load; Employing warm colors and tactile materials to enhance emotional comfort;Providing explicit visual/auditory feedback to ensure comprehensibility and operability.

### 4.2.2 Application of Affective Design in Gerontological Products

Affective design in age-friendly products manifests in specific strategies. For instance, smart home devices utilize intuitive interfaces and voice assistants to simplify environmental control, enhancing convenience and comfort. Health monitoring systems integrate reassuring visual/auditory feedback to ensure accurate health data interpretation and timely emotional support.

### 4.2.3 Practical Applications of Affective Design

This senior-oriented smart watch reduces operational complexity for elderly users through a clean and intuitive interface, large fonts, and touch feedback. At the same time, it incorporates soft color combinations and a minimalist design style to create a comfortable user experience.(see Fig. 10).

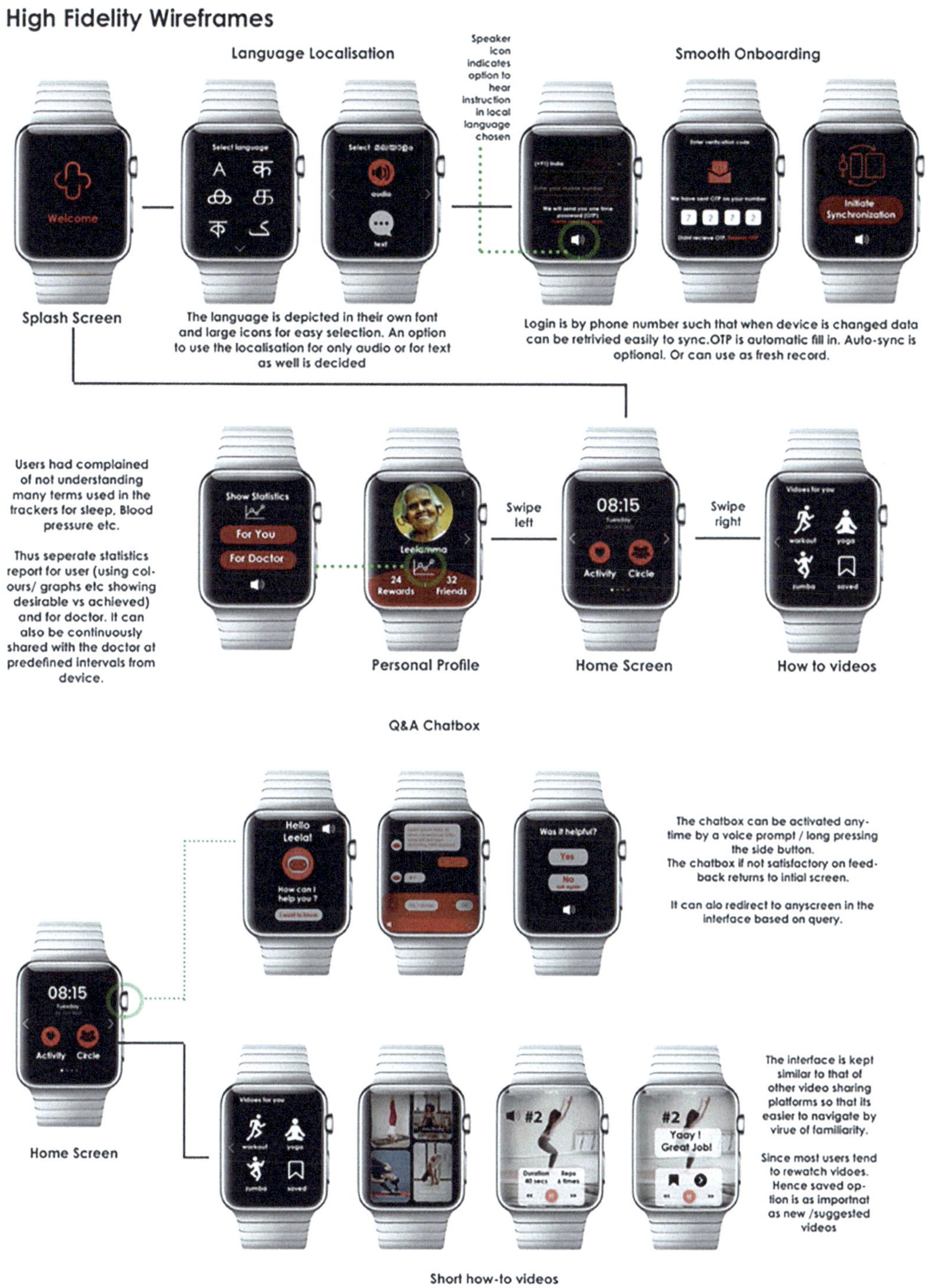


**Fig. 10** Interface design of smart watches for elderly users

# 7 Intelligent Technology and Age-Friendly Services

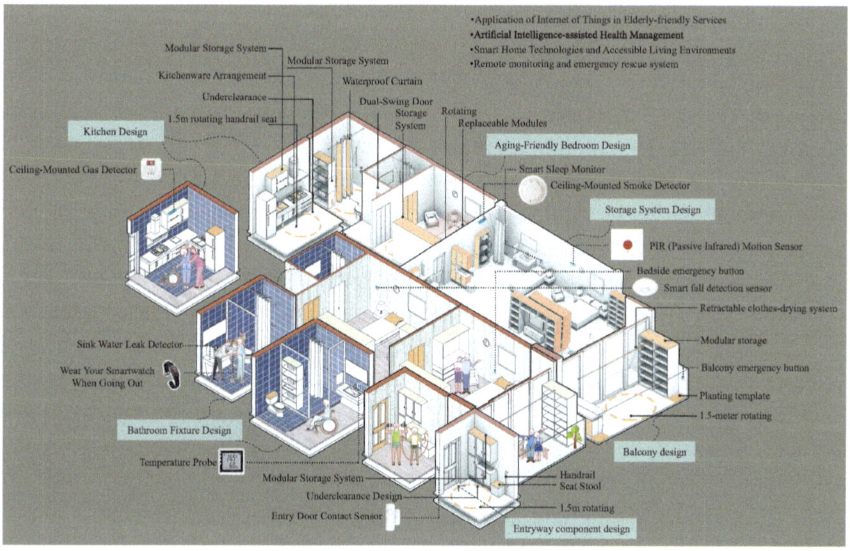


## 1 Application of Internet of Things in Elderly-Friendly Services

### 1.1 Concepts Related to the Internet of Things System

#### 1.1.1 Internet of Things Technology

The Internet of Things (IoT), originating from media, marks the third IT revolution. It connects objects via sensors and protocols through networks, enabling intelligent identification, tracking, and supervision via data exchange. Core technologies include sensors, RFID, embedded systems, AI, and nanotechnology. IoT is now applied in smart transportation, logistics, homes, healthcare, and buildings (see Fig. 1). With its broad prospects and evolving advancements, IoT promises transformative benefits for societal development.


W. Tang, *Aging-friendly Service System Design*, https://doi.org/10.1007/978-981-95-2054-1_7

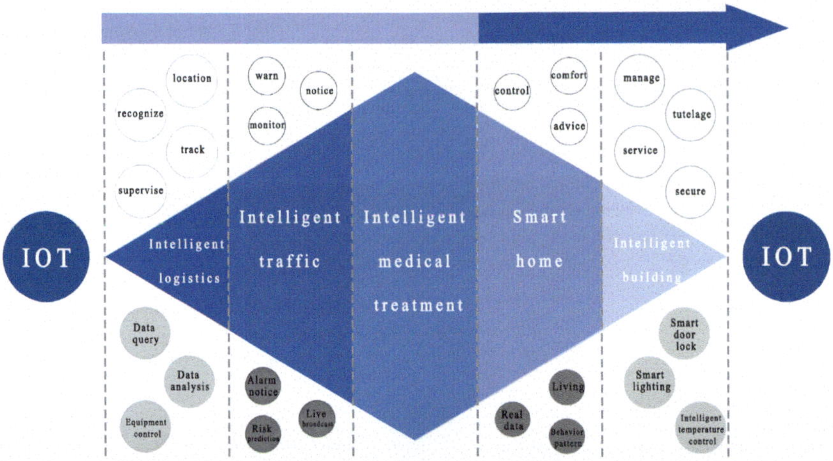


**Fig. 1** Application area of IoT technology

### 1.1.2 Internet of Things Structure

The Internet of Things (IoT) structure is typically organized into layered models, such as three-layer (perception, network, application) or five-layer, depending on classification criteria. For example, the three-layer architecture comprises the perception, network, and application layers.

The perception layer, also known as the physical layer, is the bottom layer of the Internet of Things. The main task is to collect information in the environment through various sensors, actuators and other devices, and convert this information into electronic data. These devices may include temperature sensors, humidity sensors, light sensors, pressure sensors, as well as actuating devices such as motors and relays. The perception layer is responsible for perceiving and controlling the physical world and is the basis for the Internet of Things system to obtain data.

The network layer includes various communication technologies and protocols, such as Wi-Fi, Bluetooth, Ziggbe, LoRaWAN, mobile communication networks, etc. (see Fig. 2). The network layer needs to handle issues such as network congestion, data loss, and data security to ensure the stable and reliable transportation of data. The network layer is responsible for transmitting the data collected by the perception layer to the application layer or other processing systems.

The application layer is the user interface of the Internet system, providing an interface for users to interact with the system. Its main task is to present the results of the data layer to users in an easily understandable and usable way, including various graphical user interfaces, reports, warnings and notifications, etc. Meanwhile, the application layer also provides various services, such as data query, data analysis, device control, alarm notification, etc., to meet the various needs of users.

## 1.2 The Application of Internet of Things Technology in Aging

### 1.2.1 Application of Internet of Things in Home Care

The Internet of Things (IoT) is increasingly applied in home-based elderly care, creating intelligent and personalized environments for seniors. IoT has profoundly transformed traditional elderly care models and driven comprehensive service upgrades, with applications spanning health monitoring, smart home automation, emergency response, social engagement, and data-driven personalized services.

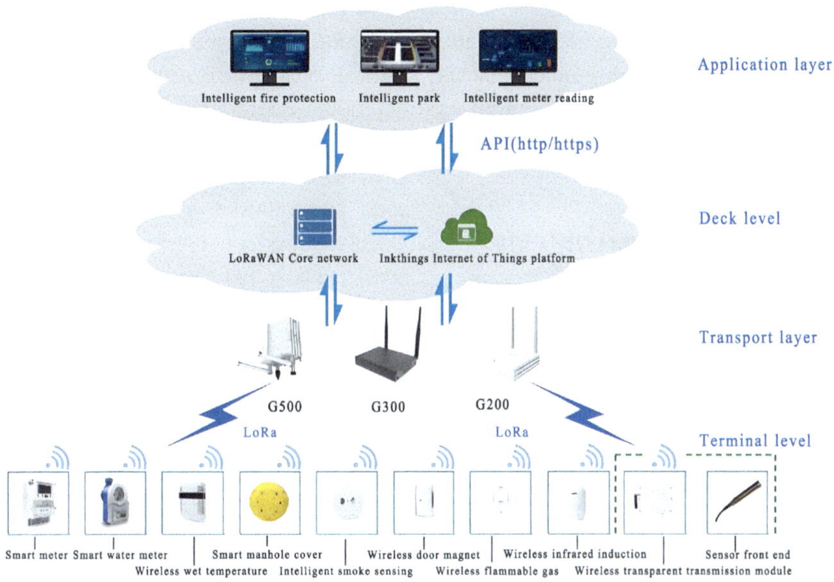


**Fig. 2** IoT and network connectivity

IoT technology significantly enhances seniors' living convenience through smart home devices like intelligent locks, lighting, and thermostats, enabling remote control via voice commands or mobile apps (see Fig. 3). This automation reduces daily burdens while enhancing overall safety and comfort. Additionally, IoT systems utilize video surveillance and security devices for real-time comprehensive monitoring, effectively and proactively preventing accidents like falls or wandering.

In emergency response, IoT technology plays a critical role. Integrated with emergency call systems and GPS positioning, IoT devices instantly transmit distress signals and precise location data during crises, significantly reducing response times and improving rescue success rates. Additionally, IoT enables data-driven

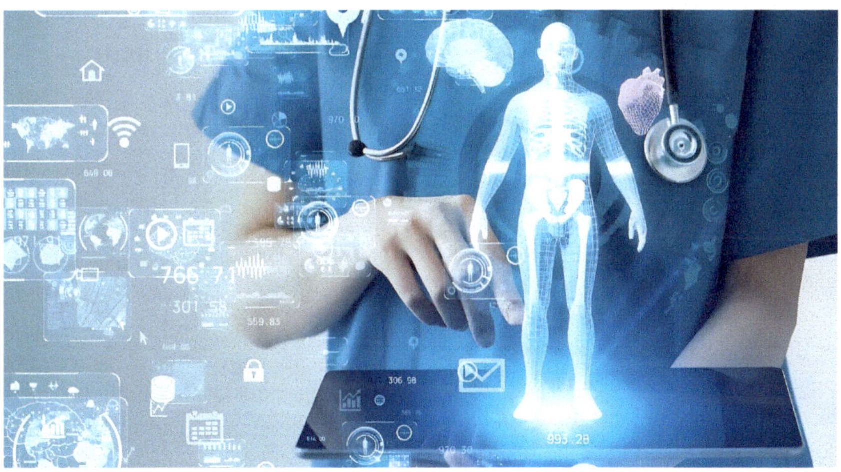

**Fig. 3** AI aided diagnosis systems

personalized care by analyzing seniors' daily habits and health needs, offering tailored health recommendations and lifestyle guidance. This approach enhances service precision, optimizes resource allocation, and ensures efficient elderly care delivery.

### 1.2.2 Application of Internet of Things in Community Pension

IoT technology integrates community-bas d elderly care resources through its extensive connectivity and intelligent sensing, creating a comprehensive platform for health management, daily assistance, and safety monitoring. Utilizing sensors, wearables, and smart home devices, the platform collects real-time physiological, behavioral, and environmental data, analyzes trends via big data analytics, and predicts health risks. This data-driven approach enables personalized care and drives the intelligent transformation of elderly care services.

In health management, IoT enables real-time monitoring and remote care for seniors. Wearables like smart wristbands and blood pressure monitors track and upload vital signs (heart rate, blood pressure, glucose) to cloud platforms, enabling caregivers or family to assess and intervene. IoT also supports remote medical consultations via video calls, providing timely professional advice and treatment plans.

In terms of life services, Internet of Things technology provides the elderly with a more convenient and comfortable home living environment. The application of smart home devices such as smart door locks, smart lighting, and smart temperature control enables the elderly to remotely control home devices through simple operations like voice or mobile phone apps. This not only enhances the self- care ability of the elderly in daily life, but also reduces the safety risks in their lives. In addition, Internet of Things technology has also achieved intelligent control over the medication management of the elderly through devices such as smart medicine boxes, effectively avoiding health problems caused by forgetting to take medicine or taking it repeatedly.

In the realm of security supervision, Internet of Things (IoT) technology has emerged as a critical enabler of proactive elderly care. By deploying interconnected smart devices such as AI-powered surveillance cameras, motion-sensitive smoke detectors, and fall detection sensors, communities establish a 24/7 intelligent monitoring ecosystem tailored to seniors' living spaces. These IoT systems continuously analyze environmental and behavioral data—detecting anomalies like sudden falls, fire hazards, or unauthorized intrusions—and instantly trigger automated emergency protocols. For instance, upon identifying a fall, the system simultaneously alerts designated caregivers, dispatches medical responders, and unlocks doors for swift access.

This seamless integration of real-time hazard identification and rapid intervention not only elevates seniors' safety and psychological well-being but also minimizes physical and logistical burdens on families and care providers. Furthermore, predictive analytics derived from IoT data enable preemptive risk mitigation, such as identifying patterns that precede accidents (see Fig. 4).

## 2 Artificial Intelligence-Assisted Health Management

### 2.1 Intelligent Medical Care and Management

In the field of artificial intelligence-assisted health management, the realization of "smart medical care and elderly care, intelligent management" is achieved by deeply integrating artificial intelligence technology, big data analysis and Internet of Things technology to build a comprehensive and personalized health management ecosystem. This system not only focuses on the health conditions of individuals, but also is committed to optimizing the allocation of medical resources, improving the efficiency and quality of medical services, thereby

**Fig. 4** Application of IoT in community aged

achieving the intelligence and smartness of health management.

### 2.1.1 Real-Time Monitoring and Early Warning

Smart healthcare systems leverage wearable devices (e.g., smartwatches, fitness bands) to continuously and accurately monitor physiological indicators such as heart rate, blood pressure, blood oxygen saturation, body temperature, and sleep quality (see Fig. 5). Data is wirelessly transmitted to cloud servers in real time, ensuring timeliness and accuracy. Cloud platforms use big data analytics to identify health trends and potential risks through in-depth data mining.

In smart management, the system integrates algorithms like machine learning and deep learning to analyze health data intelligently. Through real-time monitoring, it automatically detects abnormalities (e.g., elevated heart rate, high blood pressure, low blood oxygen) and triggers alerts via SMS, app notifications, or dashboards to users, families, or medical institutions, enabling prompt intervention.

### 2.1.2 Personalized Health Monitoring

Personalized health monitoring relies on comprehensive data collection through IoT technologies like wearables and smart sensors, which continuously track physiological metrics (heart rate, blood pressure, glucose) and behavioral patterns (activity, diet). Data is wirelessly transmitted to cloud servers, forming a personal health database that tracks real-time status and historical trends, enabling tailored insights and interventions.

The system employs big data analytics and machine learning algorithms to deeply analyze health data. By integrating multi-dimensional factors like age, gender, family history, and lifestyle, it identifies individual risk profiles and potential health issues. Leveraging medical expertise and clinical data, the system generates personalized health screening protocols that address current conditions and predict future trends, offering proactive management recommendations. An interactive feedback mechanism continuously refines detection strategies, ensuring precision and delivering tailored, efficient health management experiences.

### 2.1.3 Automatic Health Reminder and Intervention

The artificial intelligence-assisted system, relying on big data analysis, machine learning algorithm s and personalized monitoring devices, has achieved in-depth insight and precise detection of users' health status. The system continuously

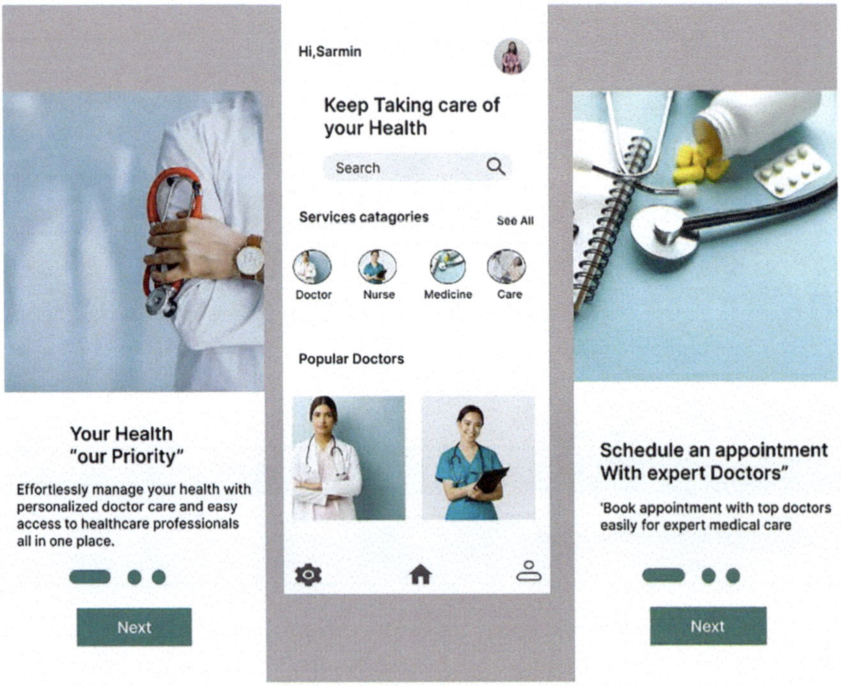


**Fig. 5** Deep mining and intelligent analysis of user health data

collects and analyzes users' physiological indicators, behavioral habits and life data, and combines users' personal characteristics and health needs to construct a unique health profile. Based on this, AI can automatically adjust the detection content and frequency, providing customized health detection plans that not only focus on the current health status but also predict future risks, ensuring that each user can receive the most suitable health management services for themselves (see Fig. 6).

## 2.2 Technological Development Trends

### 2.2.1 Future Trends of AI Technology in Health Monitoring

With technological advancements, AI will enhance precision in capturing and analyzing health data, predicting risks from subtle physiological changes. Integrated with IoT and wearables, AI enables 24/7 seamless monitoring, delivering personalized health management globally. It drives intelligent, automated health surveillance through self-learning and continuous optimization, offering precise interventions to realize the vision of "smart healthcare, intelligent management" (see Fig. 7).

### 2.2.2 AI-Driven Remote Health Monitoring Technology

AI-driven remote health monitoring relies on smart wearable devices, mobile health apps, medical sensors, and cloud-based analytics platforms. These devices continuously monitor patient vitals (e.g., heart rate, blood pressure) and transmit data to the cloud via wireless technology in real- time . Cloud platforms then apply AI algorithms to analyze the collected data (see Fig. 8).

Early Detection and Prevention: Continuous monitoring of physiological indicators (e.g., heart rate, blood pressure) enables precise anomaly detection, empowering healthcare providers to remotely adjust treatments, conduct consultations, and deliver personalized care.

Chronic Disease Management: Remote systems allow patients (e.g., diabetes, hypertension)

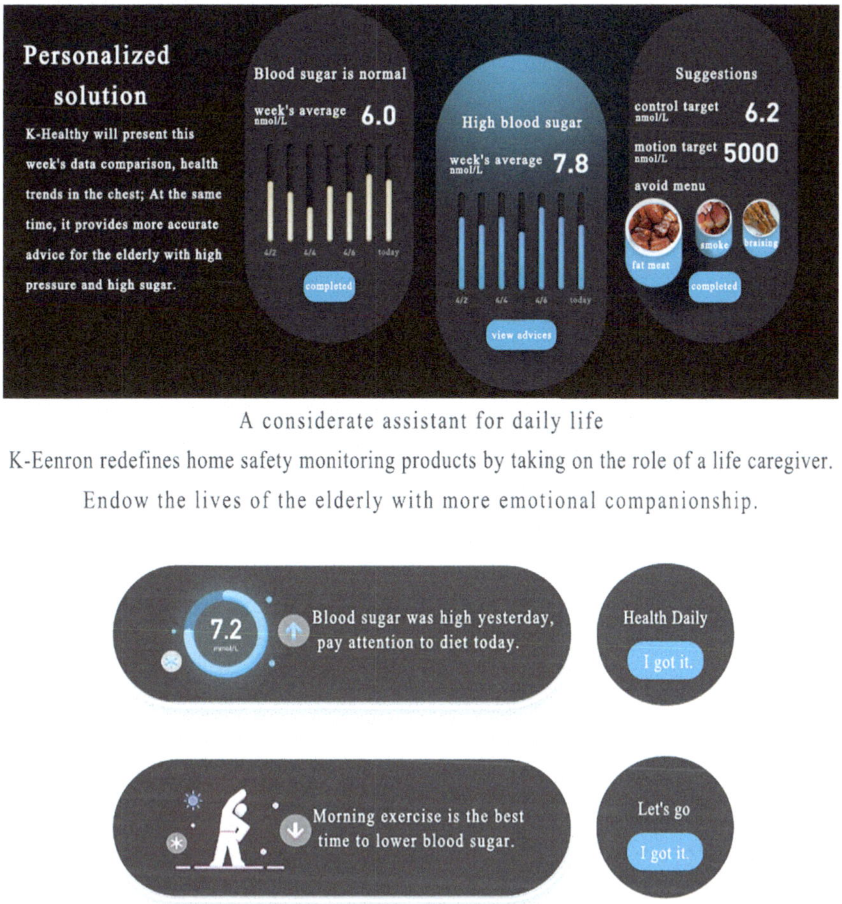


**Fig. 6** The system tailors personalized health solutions for users

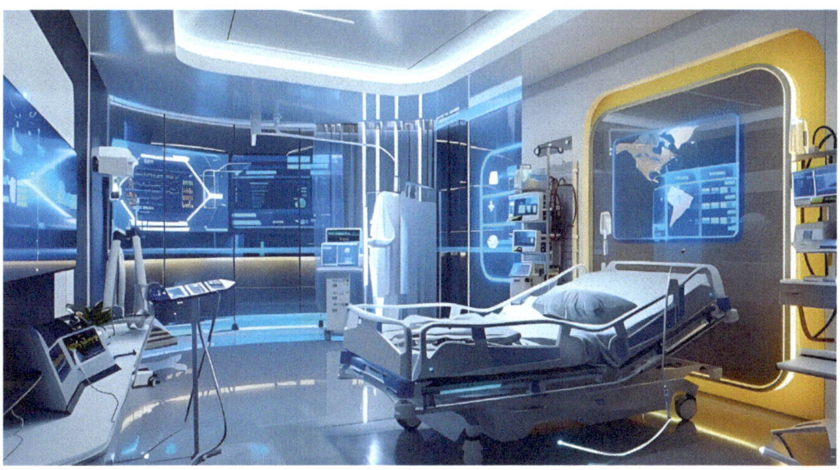

**Fig. 7** AI health monitoring

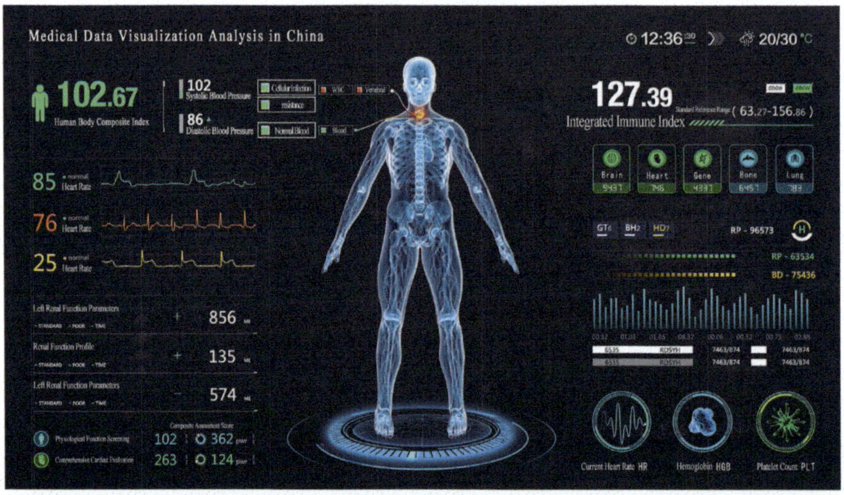


**Fig. 8** AI-driven in-depth data analysis

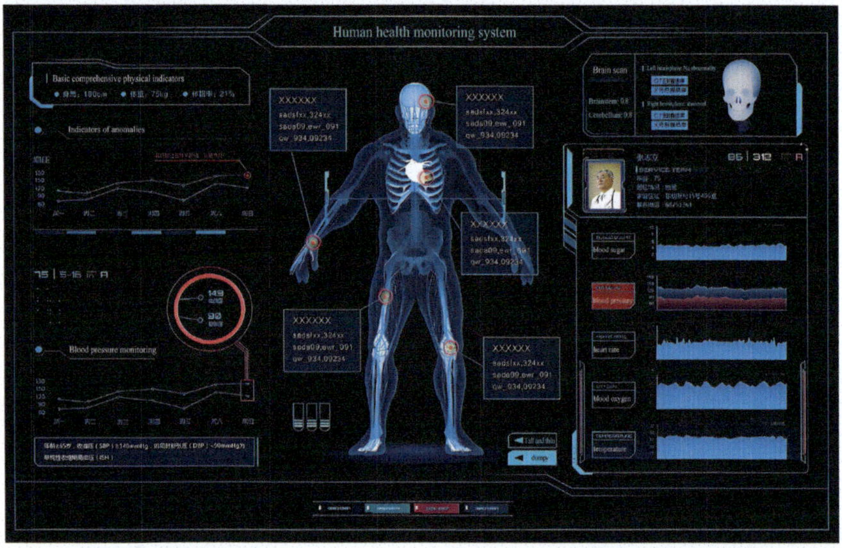


**Fig. 9** Remote health monitoring ecosystem

to track key metric s at home, reducing hospital visits (see Fig. 9). Integrated features like medication reminders, symptom tracking, and educational resources enhance self-management and preventive care.

Tele medicine Integration: Remote monitoring supports real-time health assessments and interventions, especially when in -person care is limited. Coupled with telemedicine consultations (via video/messaging), it reduces patient burdens and improves care accessibility.

## 2.3 Application Scenarios and Practical Implementations

### 2.3.1 Real-Time Health Monitoring via Smart Hardware

Smart hardware is revolutionizing health management through real-time physiological monitoring. Embedded with advanced sensors and algorithms, devices like smart wristbands and

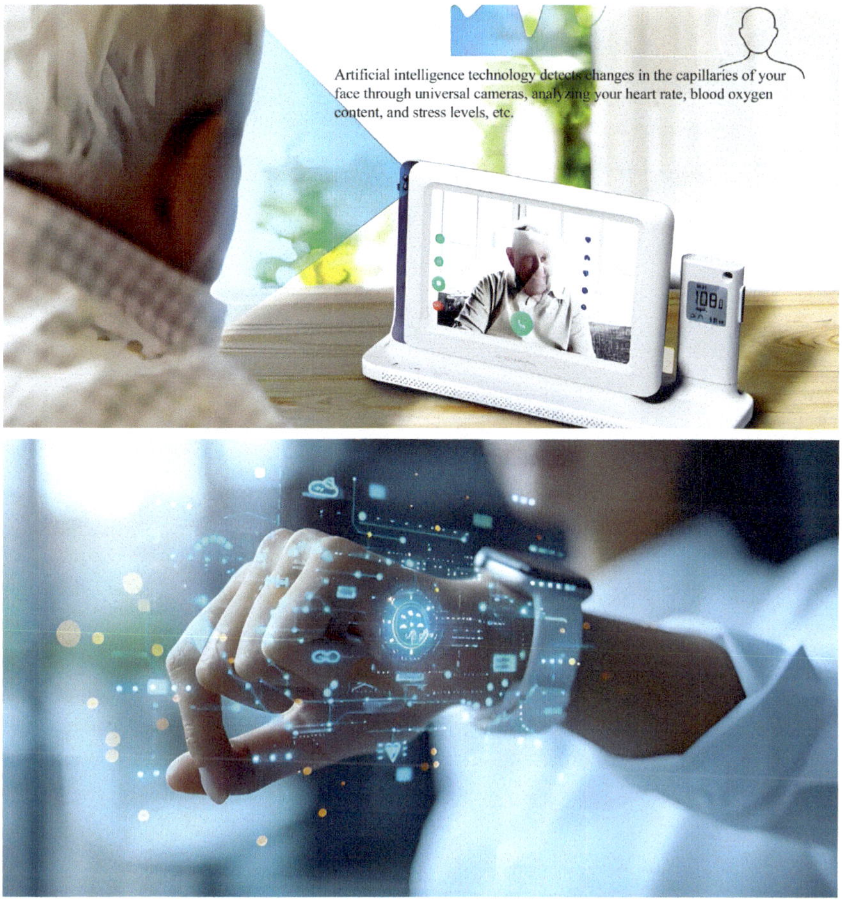


**Fig. 10** Smart hardware for real-time health monitoring

health watches continuously track vital metrics (e.g., heart rate, blood pressure, caloric expenditure), delivering precise and accessible health insights. These systems enable proactive self- care by triggering instant alerts for anomalies—for instance, detecting nocturnal cardiac irregularities and notifying users via mobile apps for timely medical intervention (see Fig. 10).

Collected data is synthesized into visualized health reports, showcasing trend analyses and historical comparisons through user- friendly interfaces. These dashboards empower individuals to track wellness patterns while providing clinicians with actionable data to optimize diagnoses and treatment plans.

By merging seamless monitoring with predictive analytics, smart hardware bridges preventive care and clinical decision-making, significantly enhancing health literacy and emergency responsiveness. This technological synergy fosters a paradigm shift from reactive healthcare to data-driven, patient-centric management ecosystems.

### 2.3.2 Generative AI for Predictive Health Risk Assessment

Generative AI demonstrates transformative potential in healthcare by excelling in predictive health risk assessment. Leveraging deep learning and multimodal data integration, these systems analyze physiological metrics, lifestyle patterns, and genetic profiles to generate personalized risk profiles. These reports identify imminent risks (e.g., cardiovascular diseases, diabetes

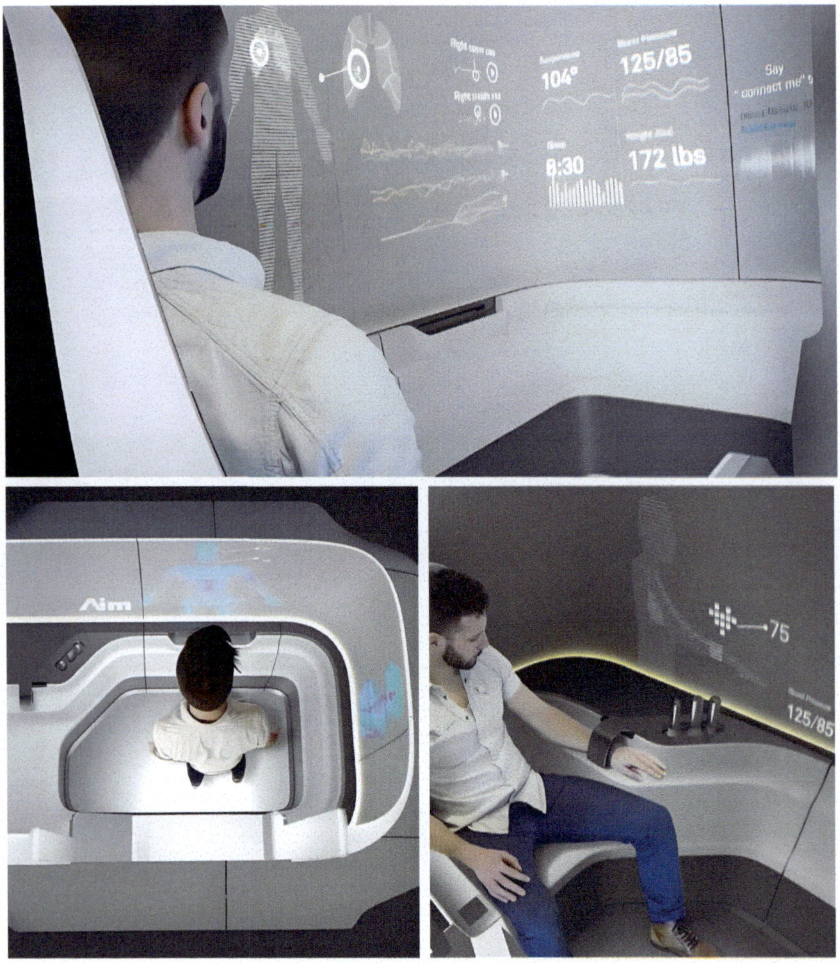


**Fig. 11** Generative AI in predictive health risk assessment

progression) and forecast long-term health trajectories through predictive analytics, enabling preemptive interventions and tailored prevention strategies.

Beyond individual care, generative AI extracts latent health correlations from population-scale datasets, informing public health initiatives. For instance, it pinpoints socioeconomic or environmental determinants of disease clusters, guiding targeted policy-making. By synergizing precision predictions with population insights, the technology enhances both clinical outcomes and systemic healthcare efficiency (see Fig. 11).

As algorithmic capabilities advance, generative AI is poised to drive intelligent, data-driven medical ecosystems, ultimately elevating global health standards through proactive and personalized care paradigms.

### 2.3.3 Design and Implementation of an AI-Driven Healthcare System for Chronic Disease Management

In contemporary society, chronic dis eases have become a severe challenge in global health, heavily burdening patients' lives and profoundly impacting the stability of healthcare systems and socioeconomic prosperity. These persistent conditions, such as hypertension (a representative cardiovascular disease), diabetes, and chronic kidney disease, are characterized by prolonged

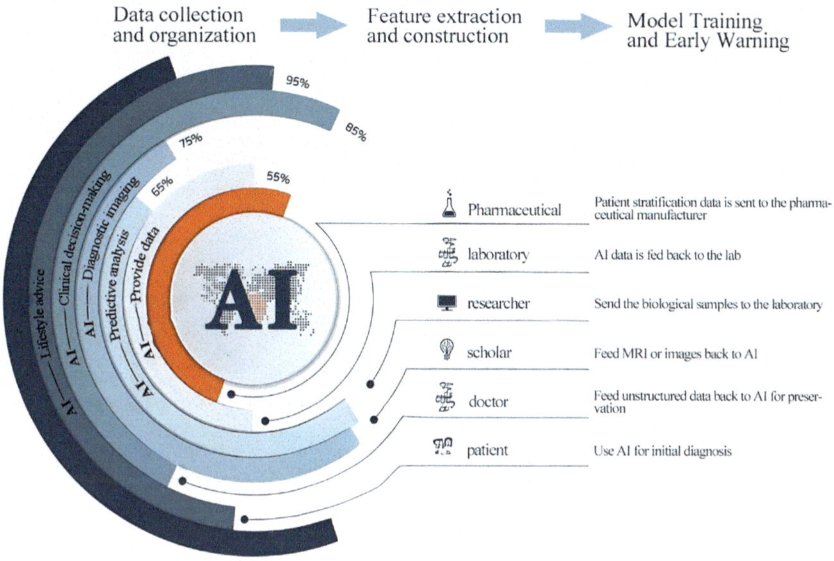


**Fig. 12** AI-driven chronic disease management system design

courses and often asymptomatic early stages, leading to frequent neglect and missed diagnosis/treatment opportunities. Current chronic disease management primarily relies on physicians' empirical knowledge and patients' self-monitoring, which faces limitations like incomplete data collection and insufficient prediction accuracy. Therefore, establishing a precise and reliable health risk warning system for chronic diseases is urgently needed. This has led to the design of an AI-based chronic disease management health system (see Fig. 12).

The system design comprises three core modules: Data Collection and Processing, Feature Extraction and Construction, and Model Training and Alerting. Through tight integration of these modules, a comprehensive health risk warning system has been developed. Rigorous testing and fine-tuning ensure alignment with practical needs. The system is deployed in medical institutions or directly serves patients, providing continuous health monitoring and personalized management support (see Fig. 13).

Data Collection and Processing Module: This module systematically collects patients' key physiological indicators, daily lifestyle information, and detailed medical histories. Raw data undergoes meticulous cleaning to remove anomalies and missing values, ensuring accuracy and completeness. Smart wearable devices (e.g., AI-powered wristbands and watches) enable real-time data capture and transmission. User-friendly application interfaces and online questionnaires further collect lifestyle data, ensuring comprehensive coverage and timely updates (see Fig. 14).

Feature Extraction and Construction Module: This module extracts predictive features from complex raw data and builds a structured feature system fr modeling. Examples include analyzing dynamic trends in blood pressure and heart rate, or exploring correlations between dietary patterns and exercise habits. These constructed features enrich model inputs and enhance risk assessment accuracy.

Model Training and Alerting Module: This module select s and optimizes machine/deep learning models, training them on historical data to achieve precise health risk predictions and timely alerts. By integrating AI algorithms, the system uses extracted features as model inputs and patients' risk levels as target labels, ensuring data-goal alignment.

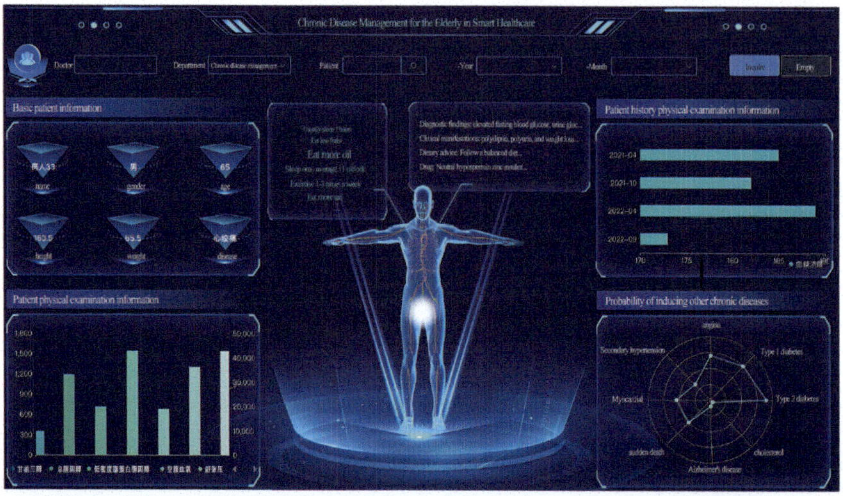


**Fig. 13** AI-integrated healthcare analysis for geriatric chronic conditions

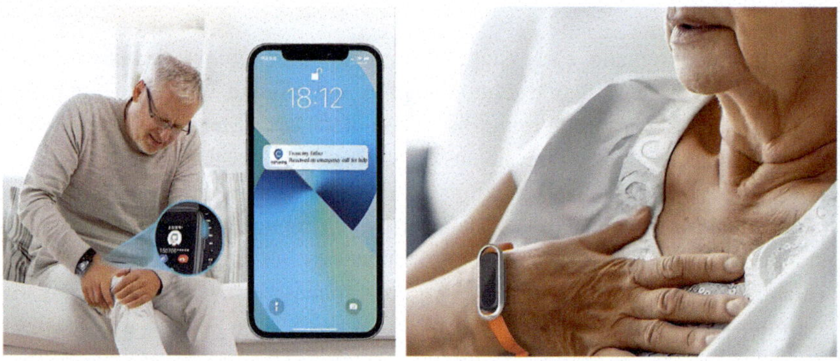


**Fig. 14** Real-time biometric capture via AI-enhanced wristbands

Integrating physiological indicators and lifestyle data allows early detection of health anomalies and customized prevention/management plans. This significantly improves healthcare precision for patients while strengthening strategic support for medical institutions and governments. Future research will focus on optimizing algorithms and models to enhance system performance and user experience, advancing the vision of a healthier, more harmonious society.

## 3 Smart Home Technologies and Accessible Living Environments

### 3.1 Core Elements of Smart Home Design

The design of smart home products carries profound significance, aiming to comprehensively address diverse group needs while significantly

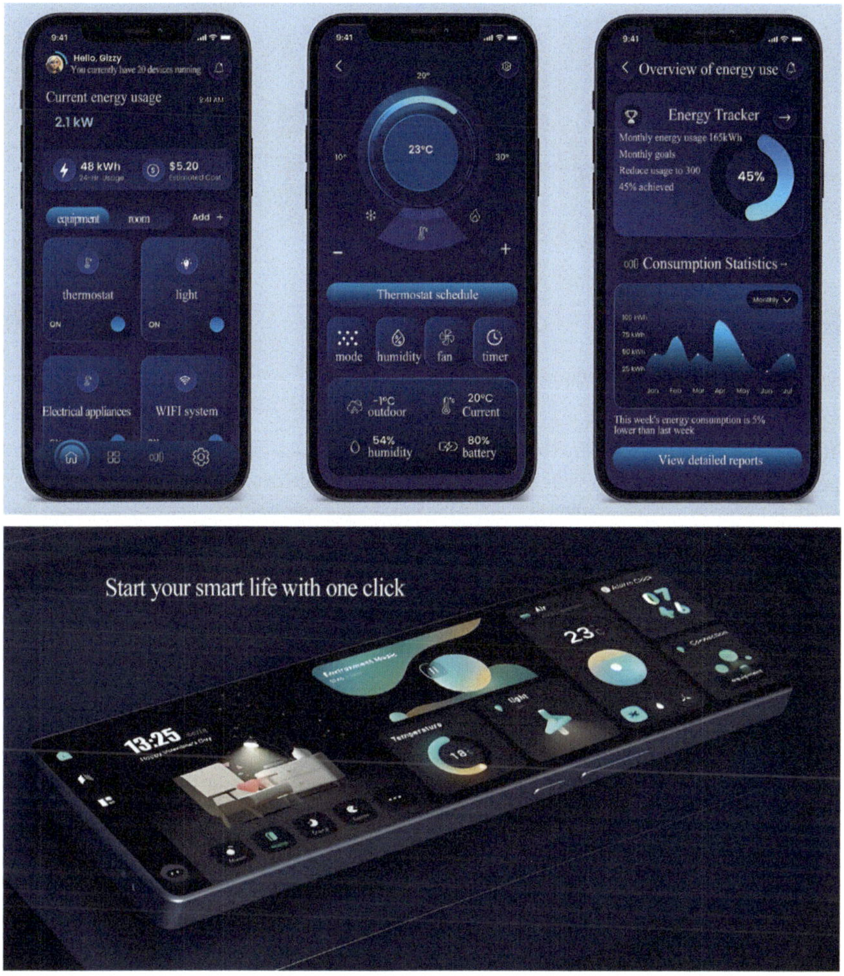


**Fig. 15** Intuitive interface design for smart home control

enhancing life satisfaction and well- being. Design concepts must adhere to core principles of usability, practicality, security, and user-centricity, ensuring alignment with real-life scenarios.

Core Elements of Smart Home Design Include: One, integration of intuitive interfaces (see Fig. 15) for elderly users' seamless adoption, leveraging IoT advancements; Two, personalized customization capabilities to deliver tailored convenience; Three, multimodal interaction design enabling fluid human-device communication via voice, gestures, or natural methods; Four, remote monitoring and device interoperability to enhance environmental intelligence; Five, prioritization of privacy protection and fail-safe mechanisms to ensure trustworthy ecosystems.

Through these measures, we strive to create comfortable, accessible, and secure living environments that actively promote seniors' physical/ mental health and overall quality of life.

### 3.1.1 Innovation-Driven Approaches in Smart Home Ecosystems

The smart home market demonstrates immense potential, with enterprises steadily penetrating this sector through diversified approaches. Current innovation-driven advancements are reshaping modern living environments across

**Fig. 16** Smart home ecosystem core components

multiple dimensions, particularly in the following key aspects:

The deep integration of artificial intelligence not only elevates device intelligence but also enables proactive learning and adaptation to user habits, delivering personalized services such as smart speakers playing preferred music and adaptive lighting adjusting brightness based on environments. In terms of security, smart homes enhance protective measures through widespread adoption of smart surveillance cameras, locks, and alarm systems, combined with remote monitoring and instant notifications to strengthen user safety (see Fig. 16).

Smart home systems prioritize energy efficiency by precisely regulating device energy consumption, exemplified by smart thermostats auto-adjusting temperatures and automated curtains managing light exposure, supporting eco-friendly lifestyles. Additionally, optimizing user experience is essential, achieved by simplifying operation processes and providing intuitive interfaces like mobile app controls or voice commands, making home management exceptionally convenient. Finally, modular de sign grants smart home systems high flexibility and customizability, allowing users to freely assemble and expand setups according to personal needs, creating tailored living spaces.

### 3.1.2 Emotion-Centric Design in Smart Home Ecosystems

Incorporating emotional considerations into smart home design not only fulfills basic functional requirements but also significantly enhances users' affective experiences during interactions. While current smart home products primarily target tech-savvy younger demographics with cutting-edge features, elderly users often struggle with operational complexity and cognitive barriers when adopting these innovations.

Resolving Core Challenges and Addressing Emotional Needs. Focusing on practicality and restoring confidence, smart devices should eliminate redundant aesthetic s and functionalities. Simplified designs alleviate seniors' anxieties about physical decline, empowering social reintegration and establishing foundations for emotionally resonant solutions (see Fig. 17).

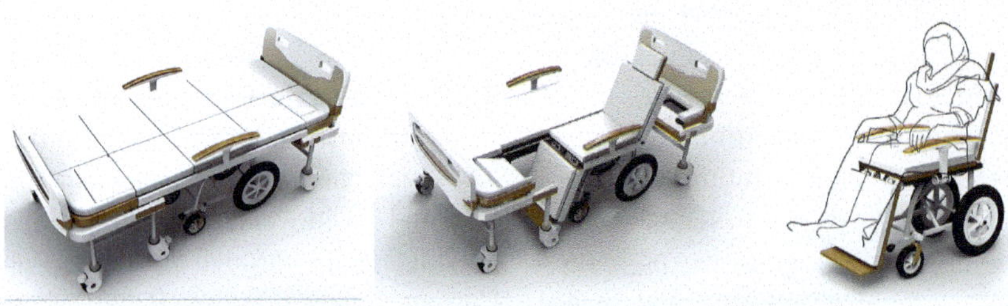

**Fig. 17** AI-optimized smart bed systems

Simplification Through Contextual Integration. Prioritizing clarity, intuitiveness, and utility reduces cognitive overload. Streamlined interfaces minimize unnecessary information and memory demands. As the critical touchpoint for human-device interaction, operational interfaces must integrate multidisciplinary design principles to ensure legibility, accessibility, and real- time feedback accuracy.

### 3.1.3 Smart Home Design with a User-Centric Experience Approach

Under the lens of user experience, smart home design aims to create an intelligent, comfortable, and convenient living environment. This approach prioritizes not only technological advancement and functional practicality but also places user needs and perceptions at its core. To build a more human-centric, convenient, and secure smart living space, the design should focus on the following key aspects.

Personalized Configuration

Smart home design emphasizes personalization and customization. By collecting and analyzing user data, the system learns and adapts to the habits and preferences of each household member, delivering tailored services. For instance, smart lighting systems can automatically adjust brightness and color temperature based on users' daily routines, creating an optimal ambiance for specific activities.

Seamless Device Connectivity

Smart home design prioritizes seamless interaction. Device connectivity should be fluid and unobstructed, allowing users to effortlessly manage and control all smart home devices without complex setup procedures, thereby realizing true intelligent living. This enables users to interact with devices through multiple convenient methods—voice commands, gesture recognition, mobile apps, etc.—ensuring easy control across diverse scenarios. Additionally, interoperability among devices is critical, ensuring systems collaborate cohesively to deliver smooth, integrated experiences (see Fig. 18).

Data Security and Privacy Protection

Smart home design underscores security and privacy safeguards. While enjoying the conveniences of smart technology, users demand robust protection for personal data and privacy. Therefore, smart home systems must employ advanced security protocols to ensure safe data transmission and storage, while transparently communicating data collection and usage practices to foster user trust.

Energy Efficiency and Environmental Sustainability

Smart home design integrates environmental and energy-saving considerations. Through

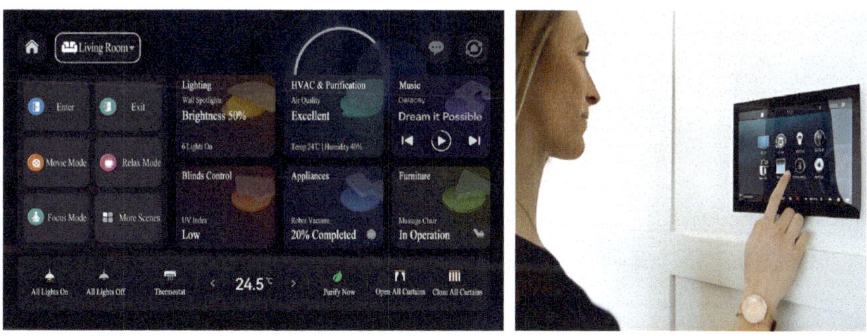


**Fig. 18** User interaction with smart devices

intelligent management, systems can autonomously optimize appliance energy consumption, minimizing waste to achieve energy conservation and emission reduction goals. This not only reduces household energy costs but also aligns with modern society's pursuit of sustainable development.

Smart home design, guided by user experience principles, is a holistic process that balances personalization, interactivity, security, privacy, and environmental sustainability. By continuously refining these dimensions, smart homes will deliver increasingly comfortable, convenient, secure, and eco-conscious living experiences.

## 3.2 Principles of Barrier-Free Design

Barrier-free design, introduced by the UN in 1974, integrates into living spaces and architecture to accommodate people with disabilities and reduced mobility (e.g., elderly). It ensures safety, accessibility, and inclusivity through adaptive solutions, addressing physical, social, and psychological needs. Combining architecture, environmental science, ergonomics, and psychology, it forms a systematic design approach. With aging populations and rising disability rates, demand grows. Future designs will prioritize universal access, smart technology, and user-friendly solutions (see Fig. 19).

### 3.2.1 Ergonomic Principles

Ergonomics focuses on studying the three fundamental elements of human, machine, and environment. In ergonomic studies, the entire system is referred to as the "human-object-environment" system, with the goal of optimizing its proportions. The application of ergonomics in design is extensive, ranging from overall spatial dimensions to specific tools within a space, all reflecting the relationship between humans, objects, and their environment. In the design of products for the elderly, ergonomics plays a significant role. For example, products like shower chairs must prioritize comfort for elderly users, ensuring the product aligns with their physical measurements. By analyzing anthropometric data of the elderly, ergonomics provides critical design parameters such as angles, lengths, and widths. It also examines behavioral patterns to determine load capacities, usage frequency, and interaction methods, offering a scientific basis for product design.

### 3.2.2 Simplification Principle

The Principle of Simplification focuses on "simplification," which means eliminating unnecessary complexity while retaining core functionality and value. This principle requires designers to consistently prioritize user needs and experiences during the design and implementation process. Through optimized design and process reengineering, products or systems achieve simplicity, ease of use, and efficiency.

**Fig. 19** Tactile paving and accessible pathways

**Fig. 20** Micro-convex faucet for the visually impaired/barrier-free water dispenser

The goal of simplification is to reduce the likelihood of user errors and eliminate operational difficulties, particularly for aging user groups, making it especially important in product design. For example, the Micro-Convex Faucet for the Visually Impaired helps visually impaired individuals determine water temperature while washing their hands. The faucet's knob features slight tactile variations that indicate temperature changes. The top of the knob is made of soft silicone. When rotated to increase water temperature, a sensor activates the silicone, causing it to protrude—the higher the temperature, the greater the protrusion. Visually impaired users can feel these changes to gauge water temperature. Another example is the Barrier-Free Water Dispenser, which provides an adaptive solution for the visually impaired. Users no longer need to align a kettle with a cup to pour water or worry about overfilling. Instead, they simply place a specially designed cup in a designated position (see Fig. 20).

### 3.2.3 Human-Centered Design Principle

The Human-Centered Design Principle focuses on "people-oriented" approaches, fostering social recognition for the elderly and facilitating their integration into modern societal development. Current mobility products for aging populations exemplify how considerate design can better serve age- friendly travel. This principle manifests in three key aspects of mobility product design: consideration, comfort, and convenience. Consideration: Clear operational steps are essential. Given the visual and auditory differences between elderly and younger users, designs should incorporate color contrast and material choices to enhance usability. For example, wayfinding systems can use distinct colors for functional interfaces, enabling modular color-coded guidance that strengthens sensory engagement and fulfills human-centered design goals. Comfort: Designs should minimize physical exertion. Features like storage baskets on walking aids address practical needs while improving comfort during travel (see Fig. 21). Convenience: Streamlined travel processes must account for elderly users' diverse needs. Innovations such as transfer reminders, boarding/ alighting alerts, and route guidance ensure products align with efficient lifestyles, ultimately enhancing quality of life.

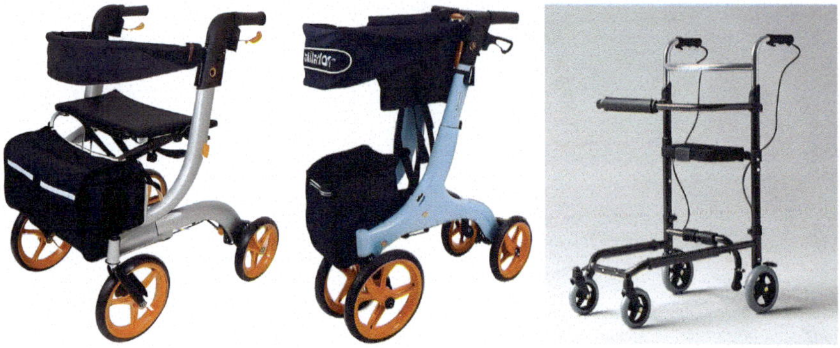

**Fig. 21** Multifunctional walker for the elderly

### 3.3 Practical Cases

#### 3.3.1 Intelligent Voice and Control System

Due to memory decline, elderly individuals often experience slower reaction times and forgetfulness. Voice recognition technology allows them to perform various tasks—such as making calls, watching TV, browsing news, and searching for information—simply by speaking. This intuitive operation simplifies processes, making technology more accessible. Integrated with big data and machine learning algorithms, smart voice systems can provide personalized health management advice, including dietary adjustment s and exercise plans, delivered through voice-guided interactions. Additionally, elderly users can control smart home devices (e.g., lights, air conditioning, TVs) via voice commands, enhancing convenience and comfort. Beyond functionality, voice assistants serve as emotional companions. Using natural language processing, they engage in conversations, alleviating loneliness and anxiety while offering emotional support and social interaction. Smart voice systems thus provide comprehensive, personalized support in health, daily living, and emotional well-being. With advancing technology and the rapid growth of smart elderly care, their applications will continue to expand (see Fig. 22).

#### 3.3.2 Visual Impairment Assistive Devices

Electronic Blind Cane aims to provide enhanced visual experiences for elderly individuals with visual impairments. The design integrates cutting-edge technologies such as computers, artificial intelligence, and cameras to enable real-time recognition and in-depth analysis of the surrounding environment. By converting captured scene information into auditory or tactile signals, it effectively assists elderly users

**Fig. 22** Intelligent voice system

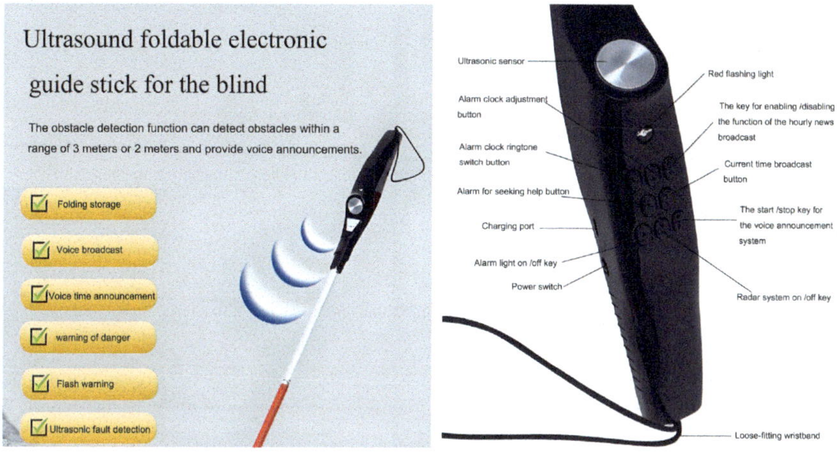


**Fig. 23** Ultrasonic foldable electronic guide cane

in identifying crowds and obstacles, improving their perception and understanding of the external environment while enhancing daily safety and convenience. The following strategies are adopted in designing such devices to ensure they meet the practical needs of visually impaired elderly users (see Fig. 23): User-Specific Adaptability: Designs must fully consider the physiological and psychological characteristics of elderly users with visual impairments to ensure comfort, ease of use, and enjoyment. The product should be easy to operate, with large, raised control buttons for tactile identification. Sensory Compensation Mechanism: When one sensory function is impaired, other senses naturally compensate to maintain comprehensive perception of external information. Designs should leverage this physiological trait by utilizing unaffected senses, such as hearing and touch, to effectively convey environmental information.

The Electronic Guide Cane is designed to provide enhanced visual experiences for elderly individuals with visual impairments. It incorporates cutting-edge technologies such as computers, artificial intelligence, and cameras to achieve real-time recognition and in-depth analysis of the surrounding environment. By converting captured environmental information into auditory or tactile signals, it effectively assists elderly users in identifying people and obstacles, enhancing their perception and understanding of the external environment, thereby improving daily safety and convenience. When designing such devices, the following strategies are followed to ensure the product meet s the actual needs of visually impaired elderly users (see Fig. 23): User Characteristics Adaptability: For elderly users with visual impairments, the design should fully consider their physiological and psychological characteristics to ensure comfort, ease of use, and enjoyment. The product should be easy to operate, with large, raised control buttons for easy tactile identification. Sensory Compensation Mechanism: When one sensory function is impaired, other senses naturally compensate to maintain comprehensive perception of external information. The design should fully consider this physiological characteristic by utilizing unaffected senses such as hearing and touch to effectively convey environmental information.

### 3.3.3 Domestic Service Robots

As multifunctional service robots emerging in recent years, household robots aim to provide diverse and convenient services. Their development has been remarkable—from Honda's ASIMO in 1997 to assistive robots in 2007, followed by SoftBank's Pepper—demonstrating not only technological leaps but also profound transformations in practical applications.

**Fig. 24** Pet robot

Currently, such robots are steadily transitioning from R&D and prototyping to small- scale commercial production, offering tangible life assistance and support, highlighting technology's positive role in enhancing elderly welfare. Today's market features various robots with diverse appearances and functions, which can be pre-programmed based on elderly users' needs. Therefore, accurately identifying elderly users' demand levels and guiding designs accordingly is crucial for reducing unnecessary costs and benefiting both businesses and consumers. Service robots that further segment user groups and clarify their demand hierarchies can provide personalized services. For example, companion robots for children should incorporate intelligent voice features (see Fig. 24), while those for the elderly require enhanced visual communication elements. Adhering to a user-need-centered design philosophy not only strengthens emotional connection and improves service experience but also avoids operational complexity and reduced usability caused by redundant functions (see Fig. 25).

## 4 Remote Monitoring and Emergency Rescue System

Intelligent tech enhances elderly care via remote monitoring and emergency systems. Sensors monitor seniors in real-time, while AI cameras detect anomalies and trigger alerts. Smartphones enable one-touch calls with GPS for quick aid. Systems prioritize data security, simplified interfaces, and user training to support seniors' safe independence and timely emergency response.

**Fig. 25** Elderly companion smart robot

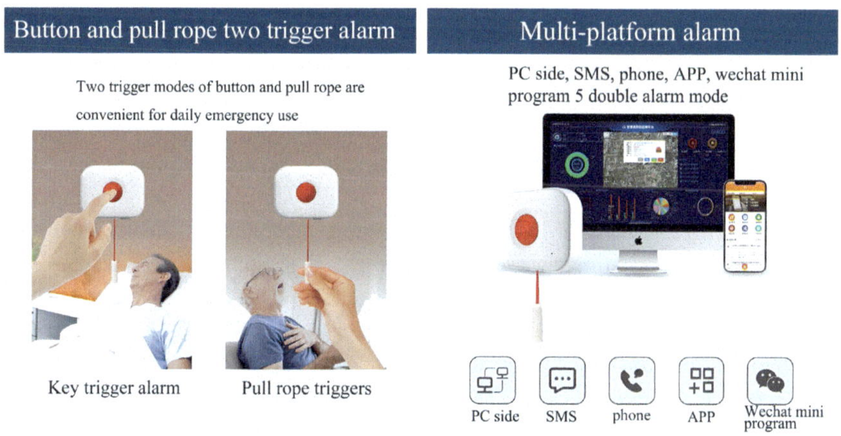


**Fig. 26** Emergency rescue

## 4.1 Requirements for Emergency Rescue Systems

The emergency rescue system is designed for senior s to quickly access assistance during emergencies. It emphasizes immediate response capability, activating rescue protocols once emergencies are detected, whether triggered manually via a one-touch call device or automatically identified by the system. Reliable communication tools, including emergency call buttons, smartphones, or dedicated terminals, ensure rapid connection to emergency contacts or professional rescue teams (see Fig. 26).

The emergency rescue system requires precise positioning, integrating GPS or other technologies to locate individuals accurately both indoors and outdoors. It should feature a multi-level alert mechanism, setting alert levels based on emergency severity and sending alerts to preset recipients like family, neighbors, community services, or medical institutions.

### 4.1.1 Real-Time Positioning and Tracking

The real-time positioning and tracking function in emergency rescue systems is a critical component for ensuring users receive prompt

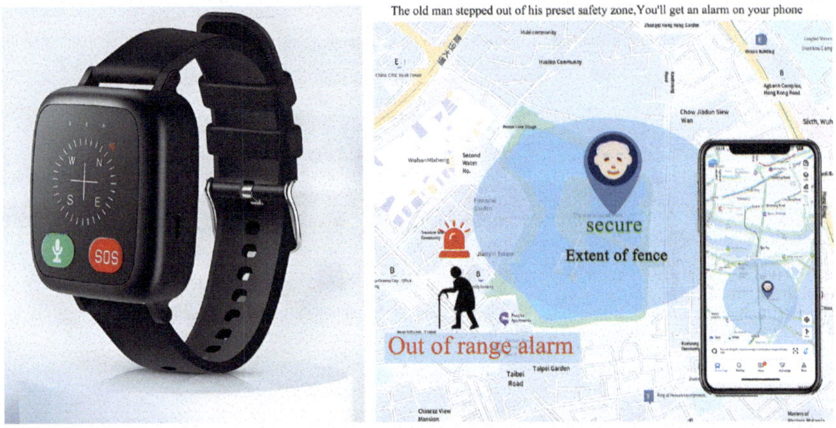


**Fig. 27** Real-time positioning and tracking

assistance during emergencies. This feature must first guarantee high-precision positioning accuracy. Whether using global navigation satellite systems (e.g., GPS, BeiDou) or indoor positioning technologies (e.g., UWB), the system should provide precise location data across environments, ensuring rescuers can quickly locate those in distress (see Fig. 27).

The real-time tracking capability of the system is a crucial function. It can continuously update location information in a dynamic environment, ensuring that no matter how the trapped person moves, the system can capture and update their position in real time. This feature is particularly important in complex and ever-changing search and rescue environments, especially when the elderly may be unable to report their locations by themselves due to accidents. The continuous tracking function has become an indispensable lifeline.

### 4.1.2 Data Transmission Reliability

In emergency rescue systems, the reliability of data transmission is critical to ensuring effective rescue operations. The application of 5G networks, with their high speed and low latency, enables rapid transmission of critical information, essential for coordinating and executing rescue efforts. 5G technology supports real-time transmission of high-definition video and large volumes of sensor data, improving the accuracy of rescue decision-making (see Fig. 28).

The integration of satellite communication technologies, such as the BeiDou system, provides reliable communication when ground infrastructure is compromised. Satellite coverage ensures connectivity in remote areas or during disasters, vital for rapid resource deployment and rescue coordination.

Data redundancy and automatic retransmission mechanisms ensure lost or corrupted data packets are promptly recovered, maintaining data integrity and accuracy. Through these measures, the system guarantees reliable delivery of critical information even in poor network conditions.

Time-Sensitive Networking (TSN) technology prioritizes critical data streams via precise clock synchronization and traffic scheduling, reducing network congestion and transmission delays. This enhances data transmission reliability in emergencies, ensuring timely delivery of rescue commands and information.

In addition, the use of encryption technology enhances the security of data transmission and protects sensitive information during transmission from unauthorized access or leakage. Constructing a comprehensive emergency communication strategy system and effectively integrating communication resources and new technologies can enhance the response speed and overall efficiency of emergency rescue.

**Fig. 28** Data transmission

## 4.2 Functions of the Remote Monitoring System

### 4.2.1 Real-Time Collection of On-Site Operation Data

One of the core functions of remote monitoring systems is real-time collection of on-site operational data. This function ensures uninterrupted reception and recording of monitored parameters, providing instant information for rescue and management operations.

The system uses efficient sensors and networks to reliably capture on- site data in varied environments. These sensors track key metrics (e.g., temperature, humidity, vibration), transmitting real-time data to centers (see Fig. 29).

The system design prioritizes data transmission stability, incorporating multiple encryption and stable transmission technologies to prevent tampering or loss during transfer. In case of network interruptions, the system enables rapid recovery to ensure continuous data transmission.

### 4.2.2 Rapid Centralized Monitoring of Data

The rapid centralized data monitoring function in remote monitoring systems is key to ensuring timely emergency detection and response. It requires systems to efficiently collect and centrally process data from various home-installed sensors, smart cameras, and devices. Whether monitoring seniors' daily activity patterns, health status changes, or identifying anomalies, the real-time nature and accuracy of data collection remain crucial.

To ensure data continuity and timeliness, the system requires robust backend support for stable operation across all network conditions. Whether via Wi-Fi, cellular networks, or Bluetooth, data must be transmitted without delay to central servers or cloud platforms. This real-time capability ensures potential risks are promptly detected and swift action is taken when needed.

The data collection process should minimize impact on users' daily lives. The system design must reduce power consumption, extend sensor and device lifespan, and ensure installations avoid disrupting seniors' routines. By leveraging advanced analytic s and algorithms, including big data analysis and machine learning, the system identifies anomalies and predicts future risk trends. This predictive analysis enables proactive measures to reduce emergency probabilities.

By meeting these requirements, the rapid centralized data monitoring function in remote monitoring systems effectively enhances seniors' safety and security levels, ensuring timely assistance when needed. This function not only

**Fig. 29** Data collection

improves their quality of life but also provides robust technical support for their health and safety (see Fig. 30).

### 4.2.3 Support Multi-terminal Access and Intelligent Analysis

The remote monitoring system supports multi-terminal access, allowing users to view monitoring data anytime, anywhere through devices like smartphones, tablets, or PCs. This flexibility significantly improves system usability, enabling users to maintain control over monitoring status whether at home, in the office, or while traveling.

The system's intelligent analysis utilizes advanced data processing and machine learning to deeply analyze large volume s of collected data. This includes real-time data analysis as well as historical trend analysis, enabling prediction of potential issue s and proposal of response strategies. Intelligent analysis enhances monitoring efficiency, reduces false alarms, and ensures alerts are triggered only when truly needed.

Through intelligent analysis, the system identifies abnormal patterns and automatically adjusts monitoring parameters or initiates preventive measures when necessary. For example, in smart elderly care, it detects reduced activity or sudden changes in seniors' routines, promptly notifying caregivers or family members for swift action. The system also offers customized, personalized monitoring solutions based on users' specific needs. Users can set alert thresholds, select data to prioritize, and define how to receive alerts and reports.

## 4.3 The Application of Remote Monitoring and Emergency Rescue Systems

The application of remote monitoring and emergency rescue systems in the field of intelligent elderly care, by integrating modern information technology, provides the elderly with a safe and convenient living environment. These systems can monitor the health conditions and activities of the elderly in real time, collect data through smart devices such as watches and sensors, and predict potential problems and propose corresponding solutions through data analysis.

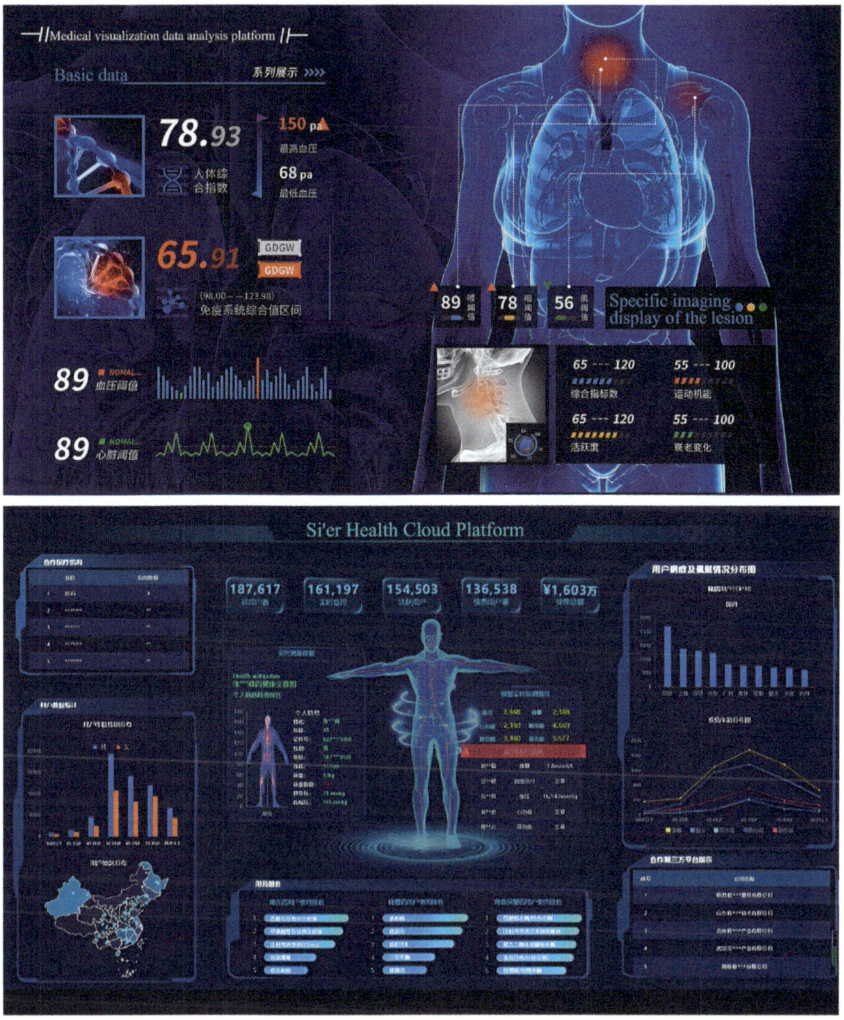


**Fig. 30** Data analysis platform

In an emergency, the system can respond promptly through a one-click emergency call button or an automatic trigger mechanism, sending the distress message to the preset emergency contact person or service center. This ability to quickly and centrally monitor data enables rescue workers to promptly understand the condition of the elderly and take corresponding rescue measures. The remote monitoring system also supports multi- terminal access. Whether it is family members, caregivers or medical service providers, they can all view the status of the elderly in real time through devices such as smartphones, tablets or computers. This multi-terminal access capability enhances the flexibility and convenience of monitoring (see Fig. 31).

### 4.3.1 Positioning and Behavior Analysis

The real-time positioning function ensures the system can quickly determine seniors' precise locations during emergencies. Whether via GPS, cellular network positioning, or Wi-Fi positioning services, these technologies provide accurate location data to help rescuers rapidly locate those in distress. The system should also include continuous tracking capability, updating location

**Fig. 31** Remote monitoring by smart phone

**Fig. 32** Elderly person's positioning watch falls and alarms

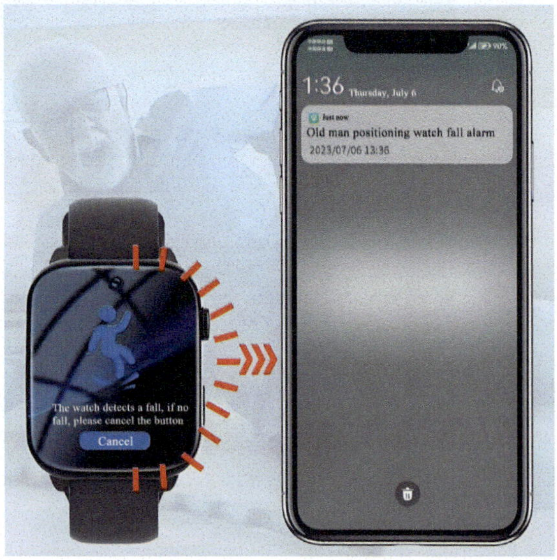


information uninterrupted in dynamic environments, ensuring real-time updates even as individuals move (see Fig. 32).

Behavioral analysis uses big data and machine learning to identify abnormal patterns and predict risks. By analyzing seniors' activity routines and health metrics, it detects early deterioration signs, alerting caregivers or triggering alarms. This predictive approach enables preventive measures, reducing emergency risks.

To ensure effective operation of these functions, system design must consider the reliability and real-time performance of data transmission, ensuring information is transmitted without delay to central servers or cloud platforms under all network conditions (see Fig. 33).

### 4.3.2 Video Surveillance and Emergency Alarm

The application of smart elderly care remote monitoring systems in video surveillance and emergency alerts is becoming a key technology to enhance seniors' quality of life and safety. By installing video monitoring devices in critical areas of care facilities or seniors' homes, these systems capture and transmit live footage to

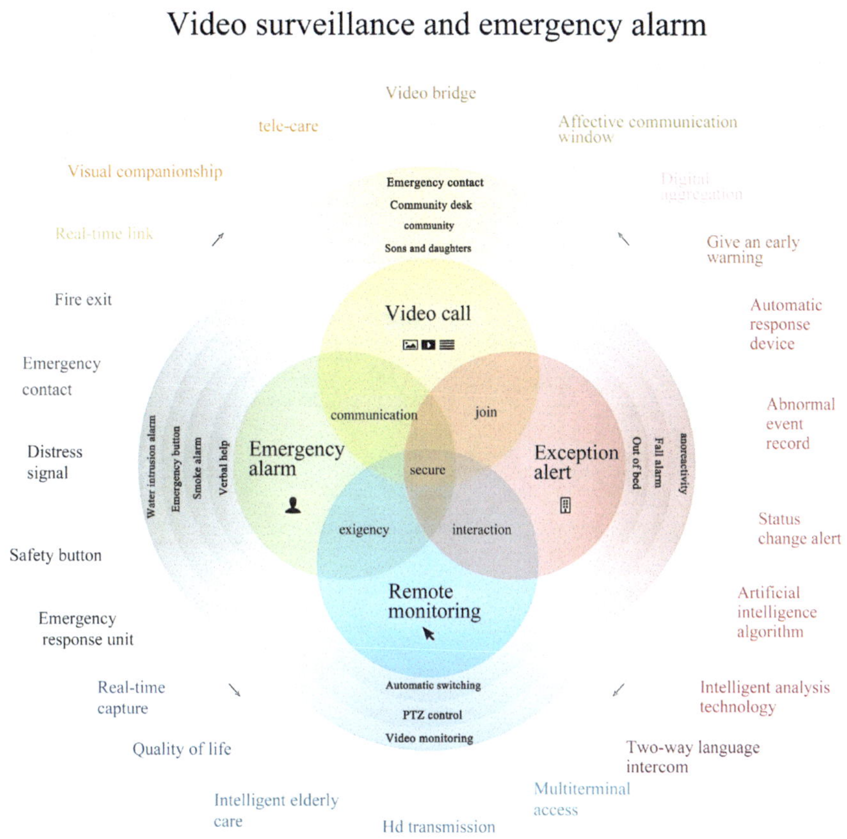


**Fig. 33** Video surveillance and emergency alarm

monitoring centers, enabling caregivers or family members to remotely observe seniors' activities and status.

Video monitoring systems typically feature high-definition transmission and multi-terminal access, allowing users to view live footage in real time via computers, smartphones, or tablets. They also support PTZ control and two-way audio communication, enabling remote users to interact with on-site personnel and intervene when necessary. Additionally, intelligent video monitoring systems integrate AI-driven analytics to detect abnormal behaviors or risks (e.g., falls, unusual activity patterns) from footage, thereby triggering alerts promptly.

The emergency alert function operates via pre-set emergency buttons or automatic triggers, immediately notifying caregivers or emergency contacts during crises. The system can also integrate with smart wearable devices (e.g., smartwatches, health monitoring bands), which automatically or manually send distress signals when seniors feel unwell or face danger (see Fig. 34).

**Fig. 34** Real-time data transmission

# 8 Evaluation and Iterative Optimization of Age-Friendly Service Systems

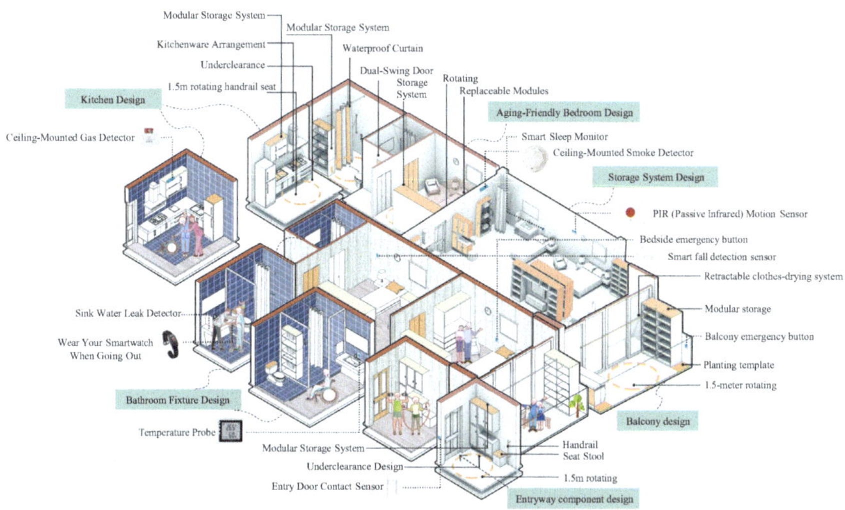


## 1 User Experience Testing Methodologies

In age-friendly system design, user experience (UX) testing constitutes a critical phase. The findings from these tests enable developers to refine the system to better align with elderly users' habits and enhance their satisfaction. As system design paradigms shift from manufacturer-centric to user-centric approaches, companies are increasingly recruiting elderly participants for questionnaires, usability testing, and focus group discussions. This participatory strategy ensures that the final designs more effectively address the specific needs of elderly populations.

### 1.1 Usability Testing

Usability testing enables direct observation of elderly users' practical operations, allowing intuitive identification of design flaws and barriers. This facilitates interface and function optimization to enhance overall user experience. The methodology delivers more direct feedback, helping designers deeply comprehend user behaviors, improve system usability, and reduce



W. Tang, *Aging-friendly Service System Design*, https://doi.org/10.1007/978-981-95-2054-1_8

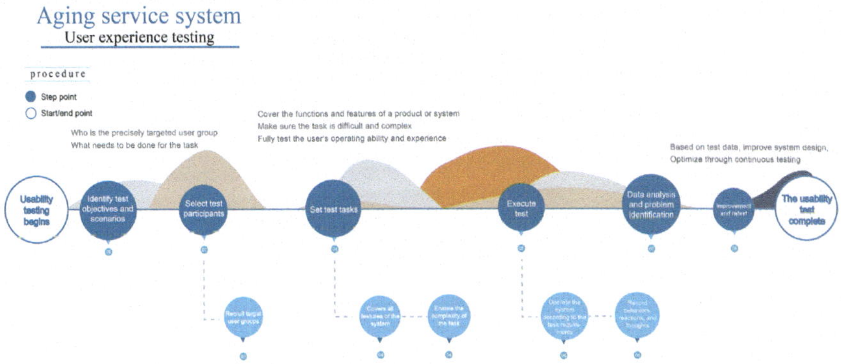


**Fig. 1** Usability testing process flowchart

elderly users' learning curve. Usability testing typically involves the following steps, as illustrated in Fig. 1:

① Define test objective s and scenarios: Clarify testing purposes and contexts, including specifications of the product/system under evaluation, target user demographics, and task requirements.
② Select test participants: Recruit representative users from target populations to ensure validity of test outcomes.
③ Develop test tasks: Design representative task sequences covering core functionalities and features of the product/system. Tasks should incorporate appropriate complexity to comprehensively assess users' operational capabilities and experiential responses.
④ Conduct testing: Guide participants through task executions while systematically documenting their behaviors, reactions, and verbalized thoughts.
⑤ Data analysis and issue identification: Synthesize test data to pinpoint operational bottlenecks and challenges encountered by participants, supplemented by collation of their feedback and improvement suggestions.
⑥ Iterative refinement and retesting: Implement design modifications based on test findings, followed by cyclical validation through repeated testing protocols. Cellular functions gradually undergo a series of physiological alterations in morphology, functionality, and metabolism as part of natural aging processes.

## 1.2 Scenario-Based Testing

Scenario-based testing evaluates system adaptability by simulating older adults' real-life contexts to capture their emotional and behavioral responses. This method reveals latent usability issues, mitigates testing bias, and generates rich qualitative data that informs design optimizations. By grounding evaluations in authentic usage environments, this approach not only strengthens system usability but also elevates user satisfaction among aging populations, enhancing their overall experience during system interaction.

## 1.3 Task Analysis

Task analysis (see Fig. 2) simulates real use cases to pinpoint system issues, measure performance, and reveal seniors' unique usage patterns. It enables tailored designs that address real user needs. By understanding user behavior, it enhances decision-making and fosters social engagement, improving seniors' lives and community.

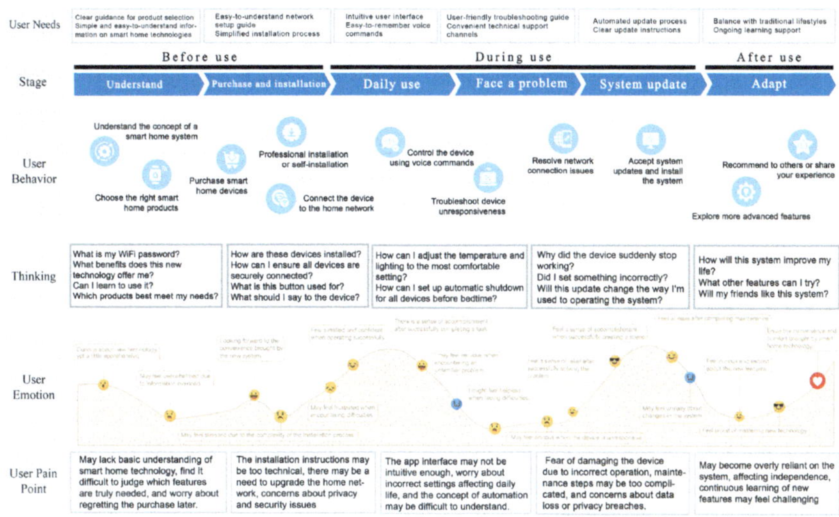


**Fig. 2** Task analysis example diagram—smart home interaction for elderly users

## 2 Usability Evaluation Metrics and Standards

### 2.1 Usability Evaluation Criteria

In age-friendly service system design, usability evaluation criteria (see Fig. 3) form the core foundation for ensuring systems meet elderly users' needs and enhance their usage experience. Based on ISO 9241-11, the evaluation criteria include: Effectiveness (ability to complete tasks successfully) Efficiency (time and effort required to complete tasks) Satisfaction (comfort level during usage).

#### 2.1.1 Framework

The usability evaluation framework for age-friendly service systems is typically based on three key elements: effectiveness, efficiency, and satisfaction. Effectiveness requires that the system assists elderly users in completing their tasks accurately and reliably. Efficiency emphasizes that users should spend less time and effort to complete tasks, with simplified processes and reduced unnecessary steps. Satisfaction means that the system interface should be clear and simple, allowing users to operate it comfortably, without confusion or stress.

#### 2.1.2 Guidelines

The design of age-friendly service systems should focus on the needs of elderly users, mainly covering interface and interaction

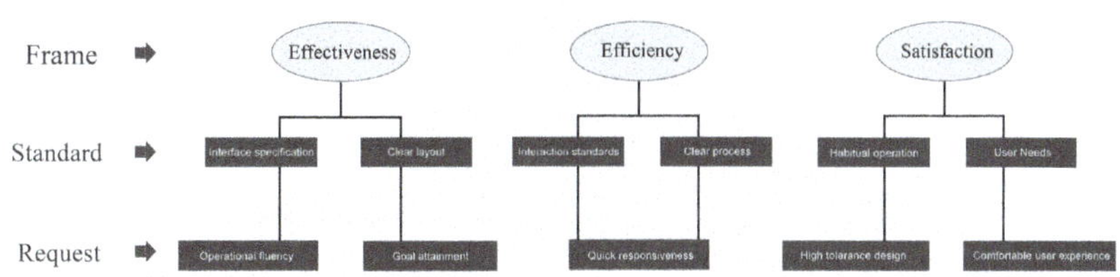


**Fig. 3** Usability evaluation criteria model

aspects. Interface design should use highly legible text, visually guided layouts, and high-contrast colors to improve readability and ease of use. Interaction design should aim to reduce cognitive load, simplify the operation process by minimizing steps, and provide clear feedback to users.

### 2.1.3 Requirements

The system requirements aim to deliver a reliable and comfortable user experience. Considering that elderly users may process information more slowly, the system should respond quickly to ensure timely feedback after each operation. High fault tolerance is necessary so that users can easily correct mistakes and face less complexity. The learning curve should be gentle, with clear, intuitive instructions and step-by-step tutorials to help elderly users quickly master the system.

## 2.2 Usability Evaluation Metrics

Usability metrics (see Fig. 4) are standardized numerical indicators used to describe the usability characteristics of a system and quantify its usability level. They translate usability issues in age-friendly service systems into specific, measurable data, such as system effectiveness, operational efficiency, and user satisfaction. These metrics provide empirical support for evaluating the user experience of such systems.

### 2.2.1 Task Completion Rate

Task completion rate is an important method for evaluating system effectiveness. A key performance indicator is the percentage of users who can independently complete assigned tasks, reflecting their ability to complete operations correctly during interaction.

### 2.2.2 Task Completion Time

Task completion time is a main method for assessing system efficiency. To ensure evaluation accuracy, a time limit is usually set for each task. If a user does not complete the task within the given time, it is considered incomplete. In addition, data such as the number of steps taken may also be recorded to provide a more comprehensive evaluation of operational efficiency.

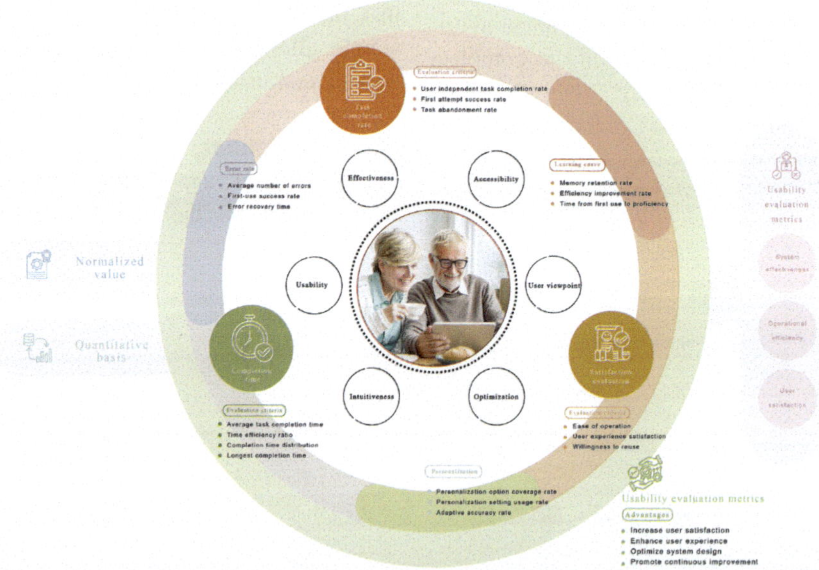


**Fig. 4** Usability evaluation metrics framework

#### 2.2.3 Satisfaction Assessment

Satisfaction assessment measures user satisfaction, although it is relatively subjective. After completing tasks, users are usually asked about the ease of operation, satisfaction with the experience, and their willingness to use the system again. Their responses are rated on a 5- to 10-point scale to make the results measurable.

## 3 Post-Usage User Feedback Collection and Analysis

### 3.1 Collecting User Feedback

Collecting user feedback enables a comprehensive understanding of the needs and usage experiences of elderly users. Common approaches include questionnaire surveys, in-depth interviews, and focus group discussions. Such feedback assists designers in identifying system shortcomings and in refining interface design, interaction processes, and functional features, thereby improving system usability and user satisfaction and better addressing the specific needs of elderly users.

#### 3.1.1 Questionnaire Surveys

In designing age-friendly service systems, questionnaire surveys (see Fig. 5) use structured forms to collect feedback from elderly users of diverse backgrounds. This method helps identify common issues, allows flexible revisions, and remains cost-effective and efficient. Anonymity encourages more genuine responses, and survey content can be easily adjusted to meet specific research needs.

#### 3.1.2 In-Depth User Interviews

In-depth interviews gather detailed insights into users' experiences, expectations, suggestions, evaluations, and specific concerns. This method reveals potential system issues and hidden user needs, offering valuable guidance for system optimization.

#### 3.1.3 Focus Group Discussions

Focus group discussions encourage participant interaction, leading to deeper insights and new ideas. They enable the simultaneous collection

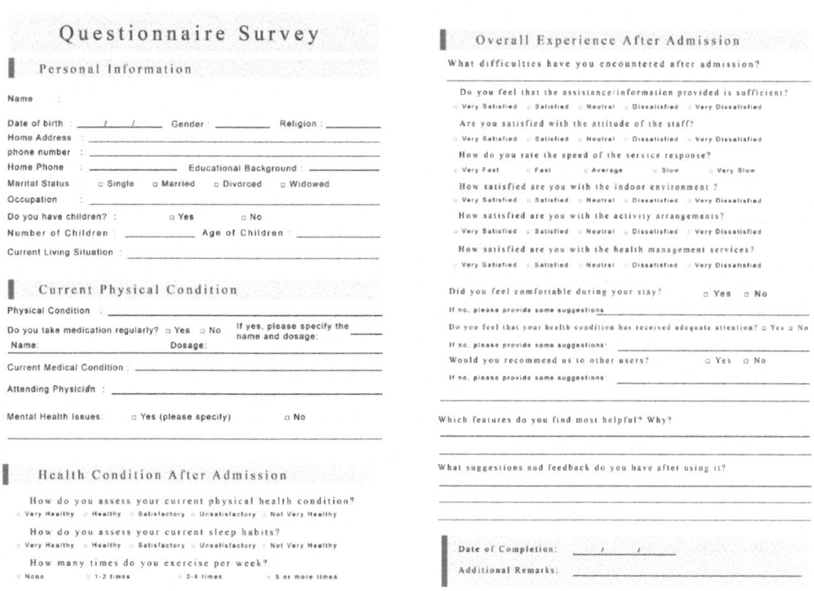

Questionnaire Survey

Personal Information

Name :

Date of birth : / / Gender : ______ Religion : ______

Home Address : ______

phone number : ______

Home Phone : ______ Educational Background : ______

Marital Status : □ Single □ Married □ Divorced □ Widowed

Occupation : ______

Do you have children? : □ Yes □ No

Number of Children : ______ Age of Children : ______

Current Living Situation : ______

Current Physical Condition

Physical Condition : ______

Do you take medication regularly? □ Yes □ No If yes, please specify the name and dosage: ______

Name: Dosage:

Current Medical Condition : ______

Attending Physician : ______

Mental Health Issues: □ Yes (please specify) □ No

Health Condition After Admission

How do you assess your current physical health condition?

□ Very Healthy □ Healthy □ Satisfactory □ Unsatisfactory □ Not Very Healthy

How do you assess your current sleep habits?

□ Very Healthy □ Healthy □ Satisfactory □ Unsatisfactory □ Not Very Healthy

How many times do you exercise per week?

□ None □ 1-2 times □ 3-4 times □ 5 or more times

Overall Experience After Admission

What difficulties have you encountered after admission?

Do you feel that the assistance/information provided is sufficient?

□ Very Satisfied □ Satisfied □ Neutral □ Dissatisfied □ Very Dissatisfied

Are you satisfied with the attitude of the staff?

□ Very Satisfied □ Satisfied □ Neutral □ Dissatisfied □ Very Dissatisfied

How do you rate the speed of the service response?

□ Very Fast □ Fast □ Average □ Slow □ Very Slow

How satisfied are you with the indoor environment ?

□ Very Satisfied □ Satisfied □ Neutral □ Dissatisfied □ Very Dissatisfied

How satisfied are you with the activity arrangements?

□ Very Satisfied □ Satisfied □ Neutral □ Dissatisfied □ Very Dissatisfied

How satisfied are you with the health management services?

□ Very Satisfied □ Satisfied □ Neutral □ Dissatisfied □ Very Dissatisfied

Did you feel comfortable during your stay? □ Yes □ No

If no, please provide some suggestions ______

Do you feel that your health condition has received adequate attention? □ Yes □ No

If no, please provide some suggestions: ______

Would you recommend us to other users? □ Yes □ No

If no, please provide some suggestions: ______

Which features do you find most helpful? Why?

What suggestions and feedback do you have after using it?

Date of Completion: / /

Additional Remarks: ______

**Fig. 5** Survey questionnaire example diagram (post-occupancy evaluation for elderly residents)

of multiple user perspectives, saving time and resources. Participants can help each other better understand system functions and offer suggestions for improvement. Focus groups also provide both qualitative data and validation for quantitative findings.

## 3.2 Data Analysis

Data analysis is a crucial step in evaluating age-friendly service systems, involving the review of user ratings and feedback to understand satisfaction and needs. Quantitative research tracks trends in scores and indexes, while qualitative analysis examines specific user opinions. Combining both methods enables designers to identify system issues and propose improvements, enhancing the experience for elderly users.

### 3.2.1 Quantitative Analysis

Quantitative analysis studies numerical data, such as elderly users' ratings, to track feedback. Key indicators include the Net Promoter Score (NPS), measuring users' willingness to recommend the system, and the Customer Satisfaction Score (CSAT), assessing overall experience. This approach offers an objective, large-scale evaluation of user satisfaction and system performance.

### 3.2.2 Qualitative Analysis

Qualitative analysis focuses on open-ended feedback, organizing user complaints, suggestions, and praise into categories like interface design and operational difficulty. It highlights common pain points, such as small buttons or complex processes, and captures subtle issues missed by quantitative methods, such as emotional responses to design elements.

### 3.2.3 Behavioral Data Analysis

Behavioral data analysis gathers users' operational data to uncover behavior patterns, common errors, and task completion rates. In age-friendly service systems (see Fig. 6), it reveals navigation habits and design shortcomings, providing valuable insights for targeted system optimization.

## 4 System Iteration and Optimization Strategies

The iterative optimization of age-friendly service systems should be user-centered, guided by user feedback and needs analysis, and

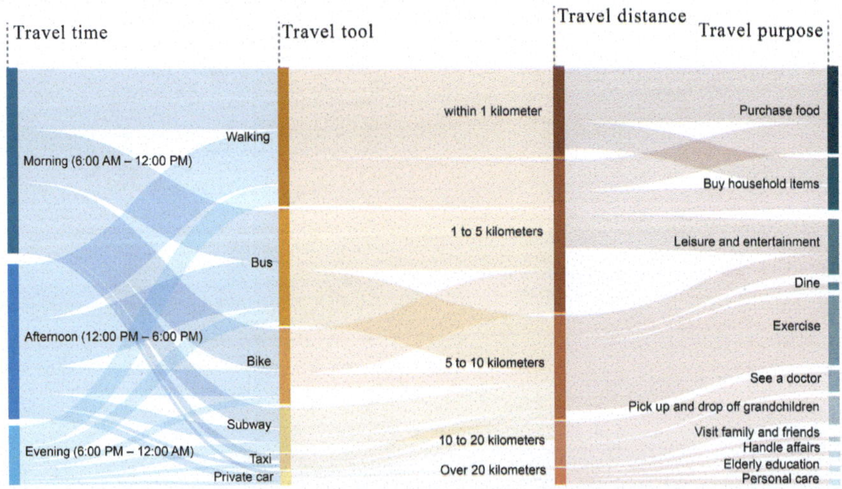


**Fig. 6** Behavioral data analytics exemplar—travel purpose patterns of older adults

focused on enhancing both form and content. This strategy not only improves user satisfaction and broadens the user base but also lowers usage barriers and promotes digital inclusion. Through continuous refinement, the system will better meet the evolving needs of the elderly population.

## 4.1 System Iteration

### 4.1.1 User Feedback-Driven Iteration Strategy

The iterative development of age- friendly service systems should be driven by user feedback, with system improvements prioritized according to user needs. A multi-channel feedback system should collect input from elderly users, their families, and caregivers. Emphasis should be placed on simplifying operations and improving interface usability, especially for frequently used but low-satisfaction functions. Key measures include streamlining processes, optimizing interface design, and introducing intelligent assistance. Through continuous iteration and data analysis, system usability can be steadily enhanced to meet elderly users' practical needs. Although the process (see Fig. 7) requires patience, it is essential for improving their quality of life.

### 4.1.2 Personalization and Adaptability Iteration Strategy

Age-friendly service systems should enhance personalization and adaptability through ongoing iteration. Intelligent algorithms can adjust interfaces and interactions based on users' habits and health conditions, such as vision, hearing, and hand dexterity. Systems should remember user preferences and offer additional personalization options, such as customized interfaces and shortcut settings, with each update to better meet elderly users' needs.

### 4.1.3 Continuous Improvement and Monitoring Mechanism

A continuous improvement and real-time monitoring mechanism should be established. Data collection systems should track key usage metrics, enabling timely detection of issues and improvements. A rapid response framework should support prompt updates and fixes. Regular security and stability assessments, along with long-term user tracking studies, will guide future iterations. In this way (see Fig. 8), the system can maintain optimal performance and adapt to evolving user needs.

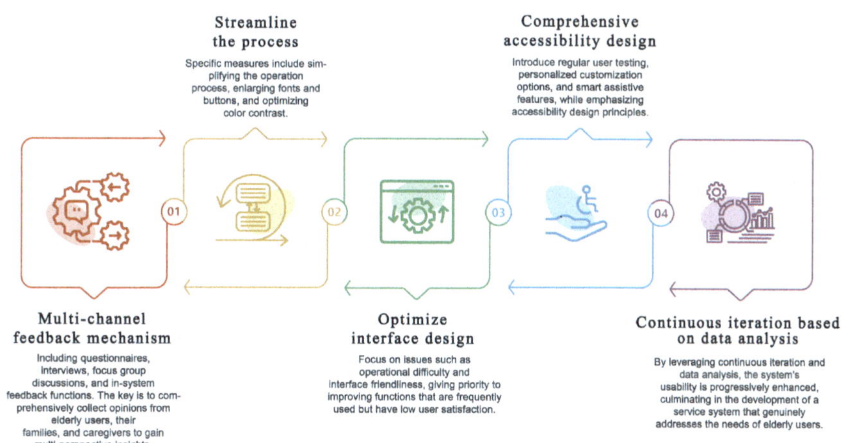


**Fig. 7** User feedback-driven iteration strategy process

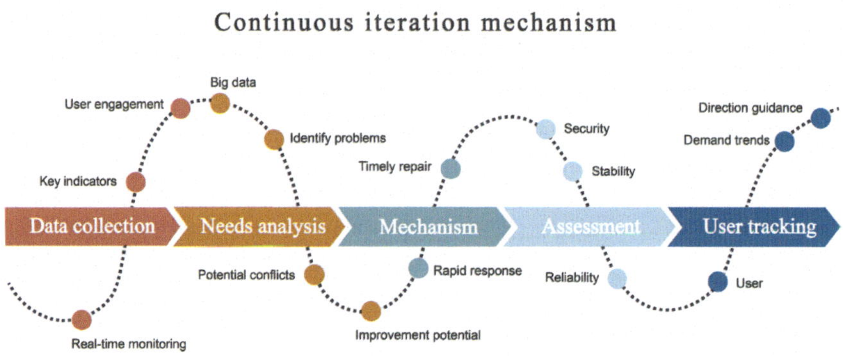


**Fig. 8** Continuous iteration mechanism

## 4.2 Optimization Strategies

### 4.2.1 Optimization of Form Elements

In the optimization design of age- friendly service systems, reasonable adjustments to layout, color, and lighting are crucial. Optimizing the layout improves elderly users' mobility and safety. Using easily recognizable colors enhances visual clarity, helping elderly users better identify their surroundings. Appropriate lighting design not only ensures spatial comfort but also increases safety.

1. Layout Planning: Age-friendly layout design should prioritize convenience and safety for elderly users. Key strategies include simplifying space circulation, ensuring smooth connections between functional areas, and avoiding complex paths and obstacles. Enhancing connectivity and flow within spaces is important. Barrier-free design should be adopted to eliminate floor height differences, facilitating wheelchair and walker access. Functional zones should be well-distributed, with public spaces centralized and stair use minimized. In critical areas, handrails, adjustable heights, and storage spaces should be added to make daily life more convenient and secure.
2. Color Usage: Elderly individuals experience reduced color discrimination, particularly with similar colors like red, green, and blue. Therefore, color optimization should balance saturation, brightness, and hue: control saturation to reduce visual fatigue, adjust brightness contrast for comfort, and use hue strategically to improve functionality and recognition.
3. A well-designed color scheme should meet elderly users' needs and, through color psychology (see Fig. 9), enhance psychological comfort, creating an appropriate atmosphere.
4. Lighting Design: Elderly individuals require higher lighting, especially for distinguishing low-contrast objects. Interior lighting should be increased by 2 to 5 times depending on the area and time of day, with larger windows for natural light and artificial lighting when needed. Overly bright single lights should be avoided to reduce glare, with light fixtures shielded. Since elderly individuals have a slower adaptation to light, lighting should focus on uniform brightness and smooth transitions, using high color-rendering light sources (see Figs. 9 and 10).

### 4.2.2 Optimization of Content Elements

When building an age-friendly service system, optimizing content elements is crucial to ensure elderly users can easily access and benefit from the service. Optimization should focus on three aspects: user experience, operational processes, and content structure (see Table 1).

1. User Experience: Focus on improving system readability and ease of use. Measures include enlarging fonts, enhancing color contrast,

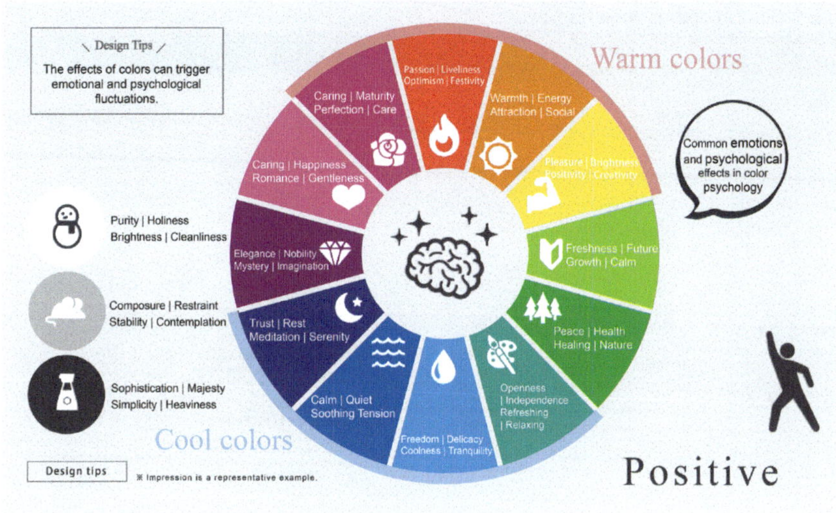


**Fig. 9** Chromo-psychological correlations—affective responses to color stimuli

| Place | General lighting standard value | Standard illuminance values for task lighting (reading, face washing, bedside) | Illuminance uniformity (UO) |
|---|---|---|---|
| Bedroom | 150lx | 300lx* | 0.60 |
| Dining room | 300lx | 500lx* | 0.70 |
| Bathroom | 200lx | 300lx* | 0.50 |
| Activity room | 500lx | 500lx | 0.70 |
| Shower room | 200lx | 200lx | 0.60 |
| Corridor | 100lx | 100lx | 0.50 |

Note: * indicates mixed lighting illuminance; the recommended illuminance for reading depends on font size, and the value provided here corresponds to medium-sized fonts.

**Fig. 10** Recommended illuminance levels for age-friendly environments

simplifying layouts, and reducing distractions. Voice feedback and control features should be added for users with visual or hearing impairments, and clear, simple language and icons should be used to ensure effective communication, improving the overall user experience.

2. Operational Processes: Simplify steps and improve efficiency by eliminating redundant processes and using intuitive interaction design. Features like zooming and button assistance should be included for user safety. Reminder functions should help with daily tasks, and personalized interface options

**Table 1** Content element optimization framework

| Content elements | Optimization goals | Optimization measures | Specific measures | Expected outcomes |
|---|---|---|---|---|
| User experience | Enhance readability and usability | Increase font size, enhance color contrast, simplify interface layout, and reduce distracting elements. | Adjust the interface font size to 18pt or above and use high-contrast color schemes. | Improve interface readability and reduce visual fatigue. |
| | Ensure clear communication | use simple and understandable language and icons | Use simple vocabulary and common symbols to replace complex terms. | Prevent difficulties in understanding and improve user comfort. |
| Operating procedures | Simplify the steps<br>Assistive features<br>Personalization | Optimize interaction | Key assist<br>Personalized settings | Improve: User efficiency<br>User satisfaction<br>User confidence |
| Content structure | Meet the daily living needs of the elderly | Organize content by categories, highlighting frequently used functions and important information. | Position frequently used features like health and finance in a visible area on the homepage. | Improve the ease of content discovery for users |
| | Provide navigation and search services | Provide clear navigation and search functions | Set a fixed navigation bar and search box at the top of each page. | Enable users to quickly locate the desired content and reduce search time. |
| | Personalized recommendations and favorites | Provide personalized recommendations and a favorites feature | Recommend relevant content based on user behavior analysis and provide a favorites feature. | Increase user content revisit rates and enhance user engagement. |

should align with elderly users' habits for an optimized experience.

3. Content Structure: Content should be organized according to elderly users' needs, highlighting frequently used functions and important information. Clear navigation and search functions should allow quick access to content, along with personalized recommendations and favorites for easy retrieval. Age-friendly content should be updated regularly for relevance.

# 9 Case Study and Future Outlook

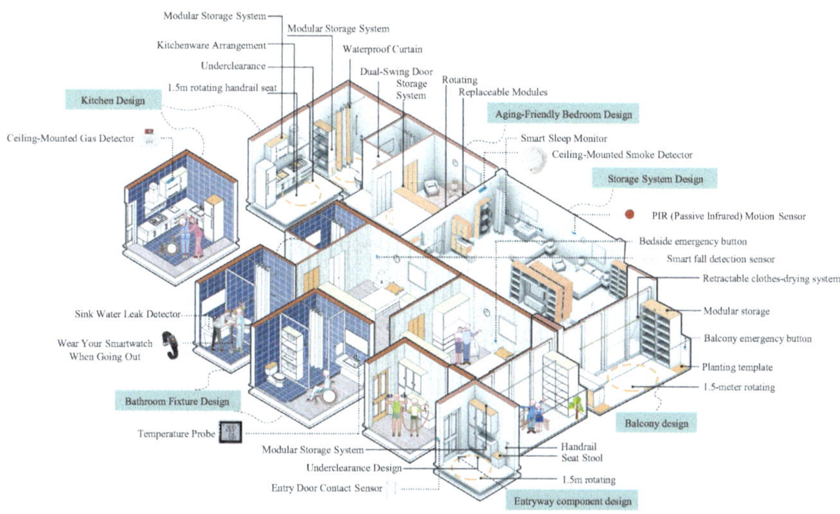


## 1 Successful Case Studies: Practices in the Design of Aging-Friendly Service Systems in China and Abroad

### 1.1 Successful Case Studies: Practices in the Design of Aging-Friendly Service Systems in China and Abroad

In recent years, the design of aging- friendly service systems has emerged as a key topic for enhancing the quality of life of older adults and promoting harmonious societal development. This chapter presents several exemplary cases from both domestic and international contexts, exploring multidimensional application models of aging-friendly services in the daily lives of elderly individuals, with the aim of offering insights for future practices in the design of aging-friendly service systems.

#### 1.1.1 Practices in the Design of Aging-Friendly Service Systems in China

1. Vanke Suiyuan Senior Living Center (Beijing)

As Vanke Group's first public–private partnership Continuing Care Retirement Community


W. Tang, *Aging-friendly Service System Design*, https://doi.org/10.1007/978-981-95-2054-1_9

**Fig. 1** Aerial view of Beijing Fangshan Vanke Suiyuan Senior Care Center

(CCRC) pilot project in Beijing, the Suiyuan Senior Living Center is committed to providing comprehensive life-care services that encompass various levels of independence, assisted living, and nursing care. This design ensures that residents can continue living comfortably in familiar surroundings even as their health or self-care capabilities change.

The center is meticulously planned with seven independent care areas (see Fig. 1) managed separately, offering 475 units and 770 beds. In addition, it features nearly 4,000 square meters of five-star public spaces, including over 20 functional areas for reading and audiovisual activities, chess and sports, calligraphy and music, dance and yoga, tea lounges, leisure meeting spaces, and private dining (see Figs. 2 and 3). Regarding its landscaping, the design takes full advantage of natural resources, integrating the living center into a park- like setting and establishing four major themed gardens: the Fitness Garden, the Dance Garden, the Wellness Garden, and the Interest Garden.

Based on extensive research involving nearly a thousand elderly individuals, Suiyuan has distilled and addressed 57 key aging-friendly needs (see Fig. 4), setting a new standard for safe and healthy senior housing. This includes systems

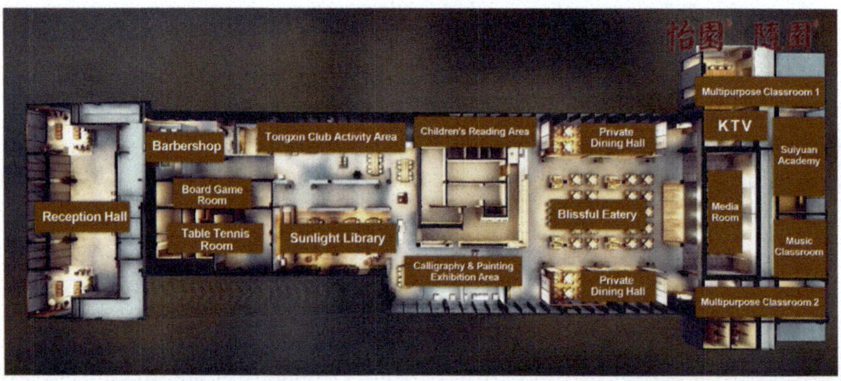


**Fig. 2** Floor plan of Beijing Fangshan Vanke Suiyuan Senior Care Center

**Fig. 3** The campus environment and public entertainment Spaces of Suiyuan Senior Care Center

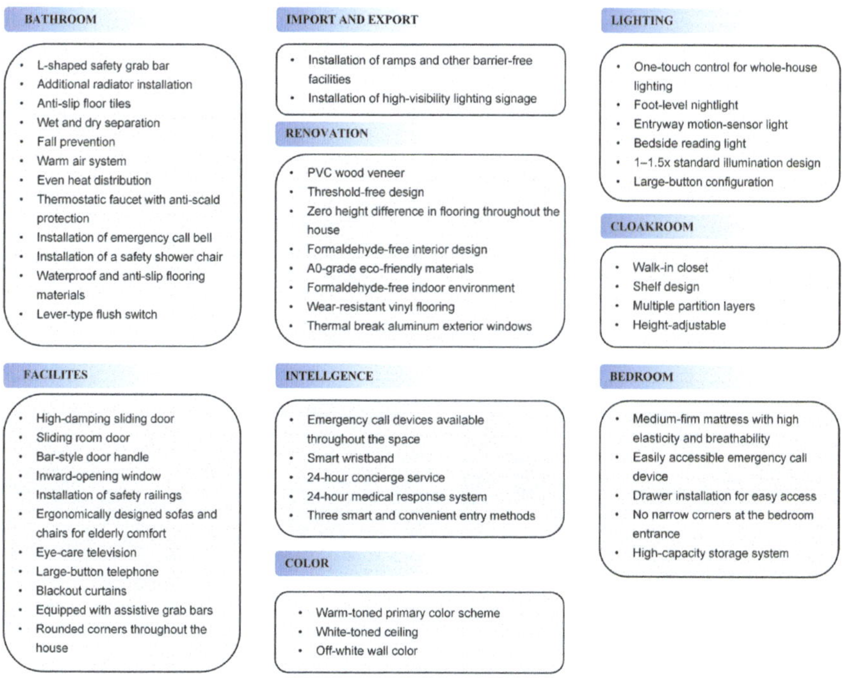


**Fig. 4** Details of the aging-friendly design of Suiyuan Senior Care Center

for lighting, sound insulation, storage, furniture, physical support, and smart home technologies (see Figs. 5 and 6).

Suiyuan actively promotes neighborly interaction through two innovative management tools: the "Happiness Bank" and "V- Care." Inspired by Japanese eldercare models, the "Happiness Bank," located in the community dining area, allows residents to accumulate points through volunteer work and participation in community activities, redeemable for various benefits (see Fig. 7).

"V-Care," developed independently by Vanke Senior Living, is a full—cycle intelligent care platform that refines eldercare service management and personalizes integrated medical- care plans (see Fig. 8). It covers eight major service modules with 165 detailed services, including

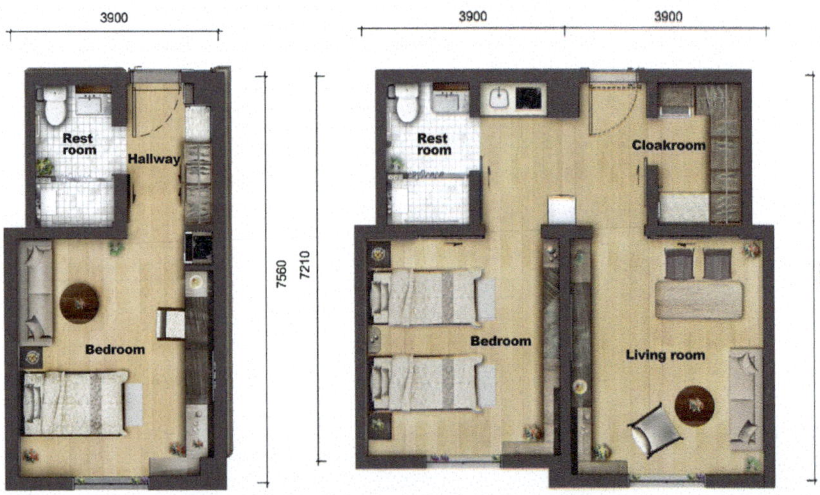


**Fig. 5** Floor plans of single-occupancy elegant residence and spacious double suite

**Fig. 6** Residential spaces in Neo-Chinese style with comprehensive furniture and appliances

**Fig. 7** The "Happiness Bank" activity in Suiyuan community

hardware support, health management, nursing care, medical services, lifestyle support, dining, and specialized housekeeping. Caregiver actions are standardized into 64 categories, and personalized care plans are tailored according to care levels, with daily care activities recorded in detail. Families can access real-time updates on the residents' health status and community activities through the platform.

The latest iteration of V-Care has introduced a cultural and recreational system, where activity managers analyze residents' engagement levels to design personalized one-on-one care plans. The system generates heat maps for community activity spaces based on attendance statistics, providing solid data for optimizing class scheduling and space utilization.

**Fig. 8** V-Care platformdesign

**Fig. 9** Aerial view of Daqing Valley

2. Taikang Community Daqing Valley (Hangzhou)

Designed by renowned Japanese architect Kengo Kuma, the Taikang Community Daqing Valley is located between the Xixi Wetland and the Zhijiang National Tourism Resort in Hangzhou. The area is surrounded by lush mountains, tea fields, and streams, making it an ideal environment for health and wellness.

The community consists of seven buildings (see Fig. 9) following the CCRC model, with independent living units and nursing apartments accommodating about 350 households. The layout extends from south to north, harmoniously blending with the natural terrain. Ten distinct garden landscapes are distributed throughout the site, creating picturesque scenes at every step. Each building is interconnected by covered corridors and underground walkways to ensure barrier-free movement (see Figs. 10 and 11).

Taikang Community adopts a "one community, one hospital" operational model and a "1+N" multidisciplinary care system, covering prevention, health management, emergency care, and rehabilitation (see Fig 9-12). The community is divided into independent living, assisted living, skilled nursing, and memory care zones, each with corresponding services to support residents throughout their aging journey (see Fig 9–14, 9–16).

**Fig. 10** Community environment of Daqing Valley

**Fig. 11** Public entertainment space of Daqing Valley

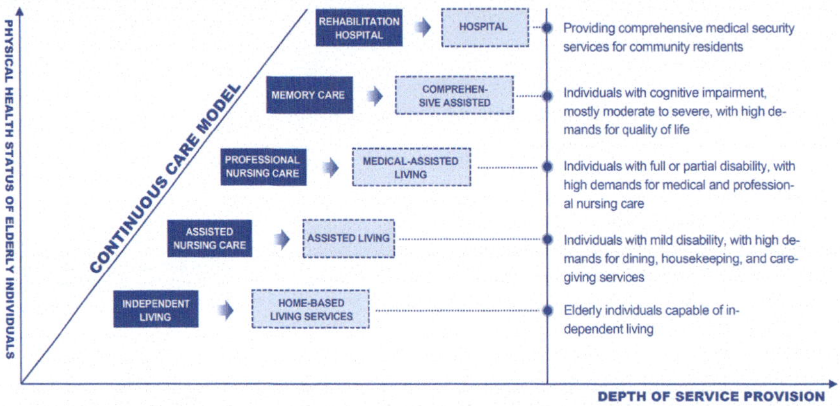


**Fig. 12** Continuous care model

Residential design extensively incorporates aging-friendly features, addressing environmental planning, interior layouts, color schemes, furniture selection, and emergency systems (see Fig. 14). Specific elements include large-font telephones, anti-glare designs, shower benches, and accessible handrails (see Fig. 13). Emergency pull cords are installed throughout the community for timely assistance (see Fig. 15).

Leveraging years of experience, Taikang has built an intelligent cloud-based eldercare platform for comprehensive lifestyle management, including clothing, food, housing, and transportation. Based on digital twin technology, the platform consists of smart care, smart security, smart concierge, and smart health subsystems, offering perception, interaction, and cognitive functions to meet high-quality living demands (Fig. 16).

Real- time data on national community deployment, resident demographics, safety, health, and lifestyle are displayed on a central digital screen (see Fig. 17). Each site is equipped with fall detection radars, security cameras, emergency pull cords, positioning mart mattresses, and vital sign radars, ensuring comprehensive resident safety.

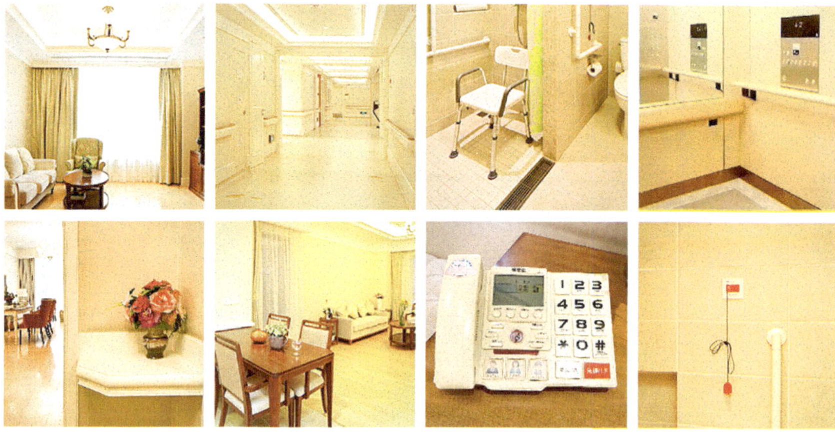

**Fig. 13** Aging-friendly and barrier-free design of Daqing Valley

**Fig. 14** Independent Living Units for self-reliant residents

**Fig. 15** Real-time health monitoring system in Daqinggu community to safeguard the well-being of seniors

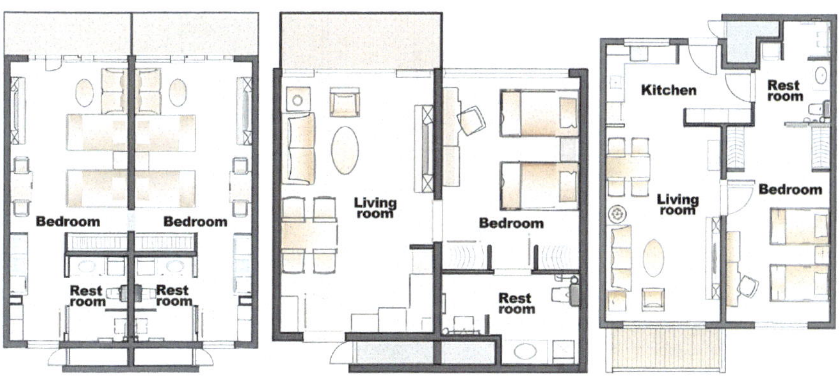


**Fig. 16** Diverse layout types of double-occupancy units in independent living buildings

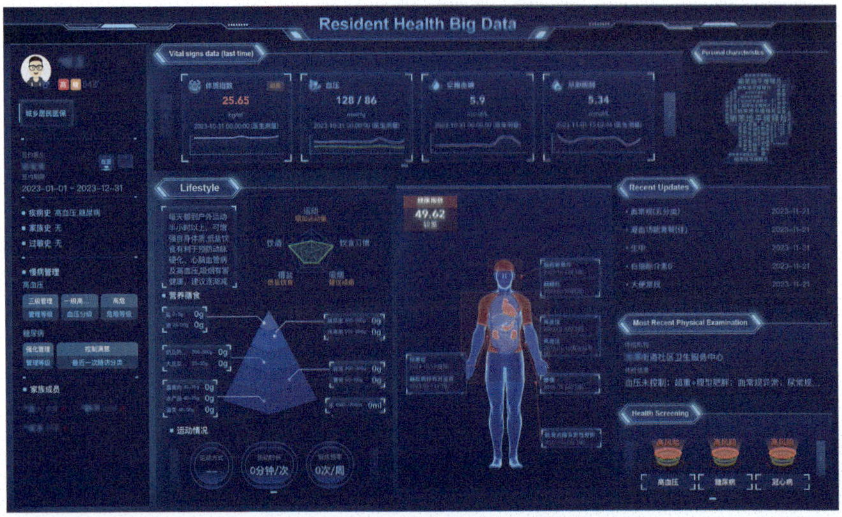


**Fig. 17** Smart community platform and digital health system of Taikang Home

3. Wendao Elderly Care Community (Shijiazhuang)

Wendao Elderly Care Community, developed by Wendao Elderly Care Group, is a demonstration project for elderly care in the Beijing-Tianjin—Hebei region. It integrates the "Wellness System, Medical-Care System, and Urban Living System" with six major supporting facilities, including a rehabilitation hospital, fitness venues, an elder university, a smart eldercare center, a dining center, and landscape gardens. The community comprises six mid-rise buildings (see Fig. 18) with a total floor area of approximately 100,000 square meters and offers over 2,000 senior beds.

Pioneering a subtropical garden- style constant-temperature senior street concept in Hebei Province, Wendao features a fully enclosed, temperature- and oxygen-controlled environment covering 3,700 square meters with a steel- frame, high-transparency glass structure. It connects seamlessly to residential and medical facilities, ensuring residents are shielded from air pollution. The temperature is maintained between 22–28 °C year-round, and humidity between 40–60%. Nearly 100 subtropical plant species

**Fig. 18** Aerial view of Wendao elderly care center

have been introduced, creating a vibrant, evergreen living environment (see Figs. 20 and 21).

The community includes ten functional zones such as a nutrition dining hall, a western-style café, a cultural theatre, yoga and dance areas, handicraft studios, reading rooms, tea houses, badminton courts, and extensive subtropical landscaping (see Fig. 19), comprehensively meeting residents' daily needs.

Wendao has advanced its smart transformation through the deployment of intelligent sleep monitors, one-touch emergency calls, smart card systems, fall detection radars, sedentary reminders, smart water meters, and an integrated control center. Its comprehensive use of face and fingerprint recognition technologies, widespread Wi Fi coverage, fall-prevention sensors, and GPS-equipped smart wristbands (see Fig. 27)

**Fig. 19** Public space design of Wendao elderly care center

**Fig. 20** Approx. 3,700 $m^2$ thermostable subtropical landscape garden

**Fig. 21** Interior layout of Wendao elderly care center

ensures resident safety. Furthermore, a full smart home system enhances living experiences via IoT technology, automatically adjusting lighting and curtains upon door unlocking (see Figs. 22, 23, 24, 25 and 26).

4. Chunxuanmao Senior Apartment (Guangzhou)

Located on Kel in Road, the Guangzhou Chun xuan mao Senior Apartment is an integrated eldercare institution that aims to create a comfortable and worry-free living environment for seniors. Based on residents' physical conditions and self- care abilities, the facility implements a zoning management strategy, offering independent living, assisted living, and dementia care units. Covering nearly 20,000 square meters (see Fig. 30), it enjoys an excellent location with convenient transportation, surrounded by mountains and parks such as South China Botanical

**Fig. 22** Warm and comfortable residential environment

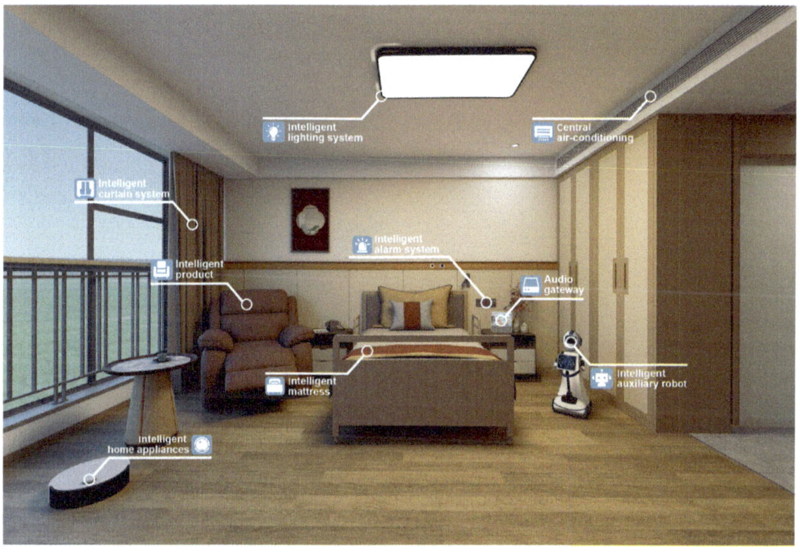


**Fig. 23** Age-friendly smart safety home systems

Garden, Qishan Park, Yushu Park, and Janta Mountain Park— all within a six-kilometer radius. These lush natural settings provide fresh air and beautiful scenery, creating an ideal wellness retreat for seniors (Fig. 27).

Additionally, several tertiary hospitals are located within a 15-minute drive, ensuring timely medical support. The nearly 10,000-square-meter amenities inside the apartment include reading rooms, chess and card rooms, pottery and painting areas, tea houses, a water bar, and multifunctional activity rooms (see Fig. 28), comprehensively meeting seniors' social and educational needs, and striving to deliver a high-quality eldercare experience.

Chunxuanmao has partnered with the leading U.S. elder care service provider Meridian to introduce internationally advanced dementia care concepts, including validation therapy, music therapy, and horticultural therapy, while integrating Chinese cultural elements. Together, they developed the specialized dementia care program called "Memory Lane." The core feature of this program is the creation of era-specific living scenes to help residents delay memory decline effectively (see Fig. 29).

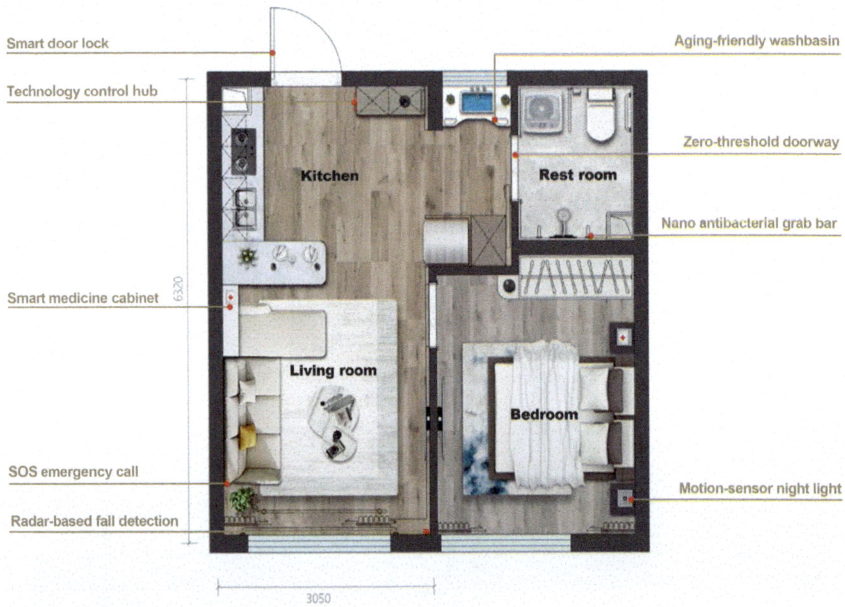


**Fig. 24** One-bedroom unit with living room and bathroom, approx. 56 $m^2$

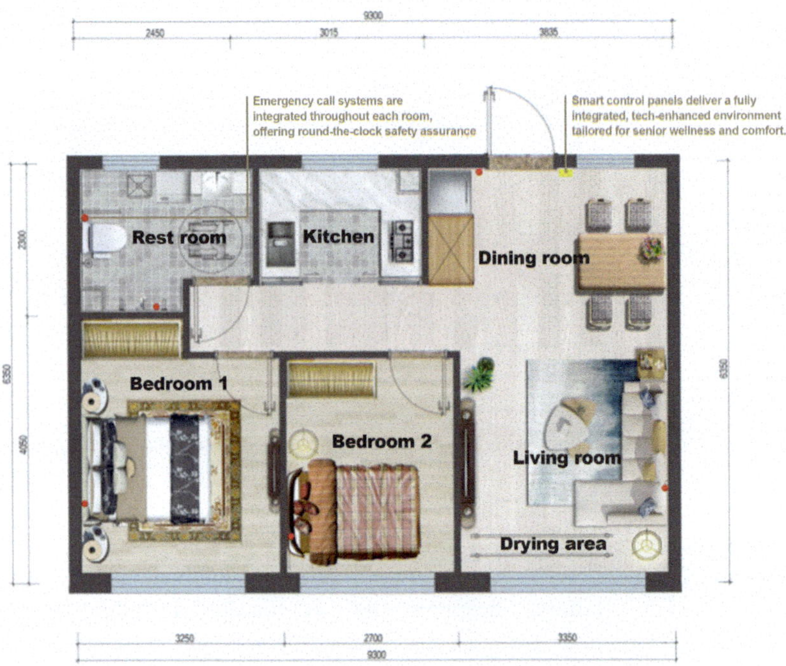


**Fig. 25** Two-bedroom unit with living and dining areas, one bathroom, approx. 82 $m^2$

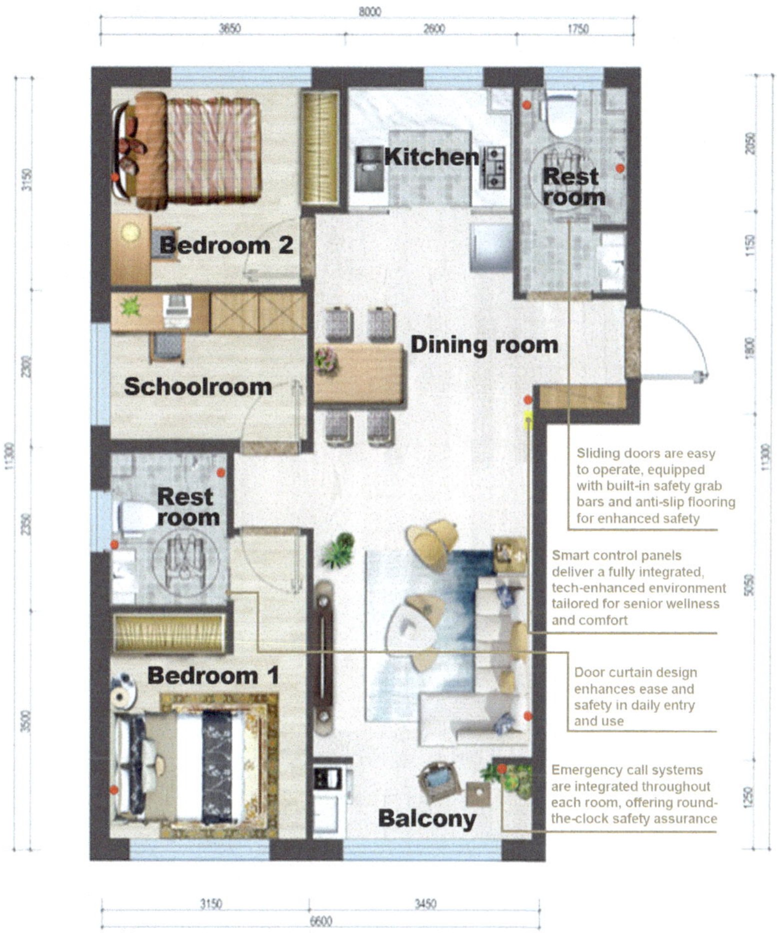


**Fig. 26** Three-bedroom unit with two living areas and two bathrooms, approx. 117 $m^2$

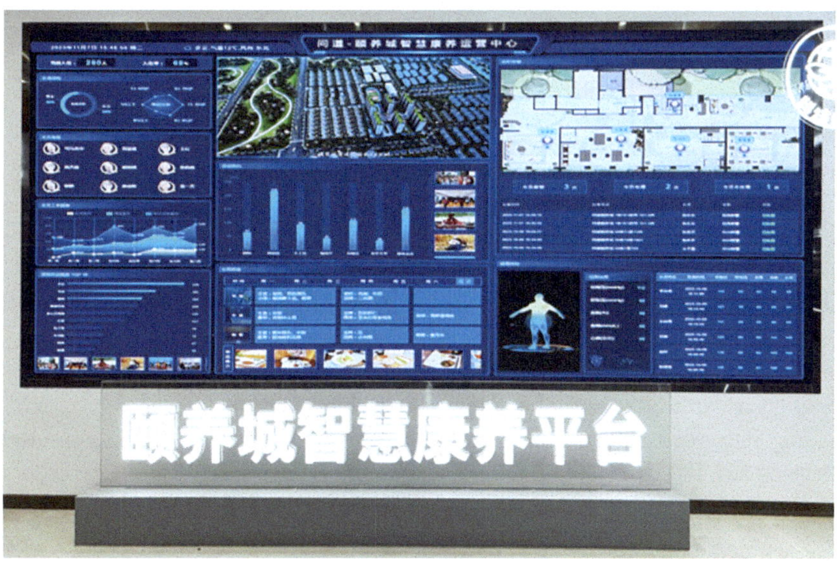


**Fig. 27** Intelligent health platform enables automatic emergency contact with medical staff

**Fig. 28** Public activity spaces in Chunxuanmao senior apartment

**Fig. 29** Memory-friendly nostalgic scenes designed to support residents with dementia

Building on mature international elder care models, Chun xanmao places strong emphasis on the professionalization of its services, operations, and management. It has developed the "Enjoy 365" service system to offer seniors comprehensive and high- quality lifestyle services, fostering a family-like community atmosphere. Personalized, professional, and technology-driven services are provided to maximize customer satisfaction and deliver internationally aligned senior care. A professional medical team of doctors and nurses is available to build comprehensive health profiles for residents, providing integrated health management services focusing on risk prevention, chronic disease management, and health maintenance by monitoring daily health data and offering timely interventions (Fig. 30).

Chunxuanmao offers a variety of room types to meet the diverse residential needs of seniors, including double rooms single rooms, small suites, and large suites, ranging from 16 to 77 square meters. Whether preferring a spacious or a compact living space, every senior can find a suitable home. Each unit comes with a spacious balcony ensuring abundant natural light, allowing residents to bask in sunlight even indoors. Rooms are fully equipped with essentials such as tables, chairs, televisions, air conditioning, and emergency call systems, ensuring a comfortable and convenient living experience (see Fig. 31).

**Fig. 30** Exterior view of Chunxuanmao senior apartment

**Fig. 31** Multiple unit types with customized padded furniture of appropriate firmness; hotel-standard comfort and cleanliness

The outdoor environment is equally impressive, featuring gardens, a 200-m green walking path, and a mini-golf course, providing seniors with ideal spaces for recreation and moderate exercise while enjoying the beauty of nature (see Fig. 32).

5. Dajia's Home · Tianjin Tai'an Road Downtown Senior Community

Located in the historic Tai'an Road cultural district in Tianjin, Dajia's Home · Tai'an Road Downtown Senior Community was formerly known as the "Garden Building," constructed in 1934 and now a protected cultural heritage site. After meticulous restoration, it has been transformed into a high—end senior community offering a wide range of eldercare services. Covering approximately 9,000 square meters and featuring nearly 90 rooms (see Fig. 34), the project maximized the preservation of the original architectural style, creating a pioneering model that blends historical heritage with modern senior living facilities, allowing residents to experience cultural richness alongside medical care.

Situated in the heart of Heping District, the community enjoys proximity to numerous medical, commercial, and tourist facilities. In the restoration process, in-depth research on the original structure was conducted, and repairs

**Fig. 32** Outdoor landscape spaces offering leisure and social opportunities for seniors

were carried out following cultural relic restoration standards. Great efforts were made to restore the classical entrance style of the Garden Building and to preserve the central courtyard (see Figs. 33 and 34). Details like coffered ceilings and intricate carvings have been carefully recreated, reviving the essence of classical ceiling craftsmanship (see Fig. 35). Some antique furniture pieces over a century old are also featured. Without altering the original structural layout, the façade was restored, and the interior space optimized with new medical service centers, dining halls, reading rooms, and age-friendly facilities like handrails and alarms, blending functionality with heritage.

Here, seniors can revisit cherished childhood memories while immersing themselves in an atmosphere rich with history and culture, as sunlight streams through the historic building, filling it with warmth and nostalgia. see Fig. 9–34, 9–35.

The Tai'an Road community has developed a comprehensive eldercare service system covering six major domains: medical care, nursing, rehabilitation, recreation, dining, and housing. One of its highlights is the medical-eldercare integration model. Through strategic cooperation with WeDoctor Hospital, the community offers in-depth integration in health checkups, chronic disease management, and green medical access pathways. Leveraging an international digital health platform, the community fully meets residents' diverse healthcare needs, addressing their concerns for future medical care.

The community welcomes seniors requiring independent living, assisted living, or long-term care, providing a full spectrum of services focused on health, rehabilitation, warm hospitality, and professional nursing. Based on international rehabilitation concepts, the rehabilitation system aims for functional improvements that are "tangible, visible, and traceable," offering both general and specialized services in cardiopulmonary, neurological, and orthopedic rehabilitation.

Care services, precisely tailored to seniors' physical conditions, include palliative care, personalized butler services, daily living support, dementia care, restorative nursing, and chronic disease management.

Nutritional services are also a key focus, with menus designed around seasonal changes and seniors' specific health needs. Meals adhere to four principles: safety, nutritional diversity, delicious wellness, and functional nutrition, promoting scientific and healthy eating habits for seniors.

**Fig. 33** Interior retains classical architectural style; antique furniture over 100 years old

**Fig. 34** Exterior of "Home for All"—Tianjin Tai'an Road Central Senior Living Community

**Fig. 35** The atrium space of Senior Community

As for the living environment, the room design artfully combines traditional British style with Chinese elements, creating a harmonious East-meets—West living space. Multiple room types are available, all following a unified, professionally designed age-friendly standard to meet varying needs (see Fig. 36).

6. Kunming Dianchi International Health and Retirement Resort Gudian Wellness Park

Located on the southern shore of Dianchi Lake in Kunming, Yunnan Province, the Colorful Yunnan · Gudian Town project covers 16,000 acres (Fig. 37). It is a large-scale cultural tourism complex integrating cultural experience, sightseeing, leisure wellness, business conventions, entertainment, and office functions (see Fig. 38).

The project encompasses eight themed areas: Gudian Cultural Core Area, Ethnic Culture Display Zone, Ecological Landscape Demonstration Area, Modern Tourism Service Zone, Gudian Wellness Park, Colorful Yunnan Happy World Theme Park, a Civic Projects Demonstration Zone, and New South Kunming Core District.

The Gudian Wellness Park sits at the heart of the Gudian Town, adjacent to the southern shore of Dianchi Lake, covering 2.2 million square meters. Surrounded by 3,800 acres of highland lakes and lush mountains, and backed by the broader 16,000-acre cultural tourism facilities, the park aims to bring the world's top eldercare resources and concepts to the region.(see Fig 9–37)

Its planning features complete age- friendly living, healthcare, and cultural facilities (see Fig. 37), offering a full range of services in living support, mental wellness, cultural entertainment, and rehabilitation. It is an international vacation-style retirement community combining independent living, assisted living, and skilled nursing care, striving to become a global benchmark for high- standard wellness and retirement living.

The Dianchi Kangyue Hospital within the park features outpatient and inpatient departments, equipped with cutting-edge medical equipment like CT scanner s and ultrasound devices (see Fig. 39). The hospital is integrated with provincial and municipal health insurance systems, providing comprehensive services spanning prevention, treatment, nursing, rehabilitation, emergency care, health management,

**Fig. 36** Living spaces combine traditional British style with Chinese elements, following standardized age-friendly residential systems

and physical examinations—allowing residents to receive medical attention easily without leaving the community.

The rehabilitation center features five specialized rehabilitation therapy rooms, equipped with internationally advanced equipment, focusing on neurological, geriatric, orthopedic, pain, and traditional Chinese medicine rehabilitation, catering to diverse needs.

The Care Center is dedicated to serving bedridden, semi-dependent, and cognitively impaired seniors, offering integrated services covering medical care, rehabilitation, and palliative care.

The Dianchi Senior University is equipped with spacious, well-lit classrooms and a variety of facilities, including clubs, a small theater, a library, a fitness room, ping-pong rooms, billiard rooms, IT classrooms, handicraft rooms, calligraphy rooms, Western painting studios, dance rooms, vocal music rooms, and musical instrument rooms (see Fig. 40). It accommodates over 500 seniors for daily activities, making it an essential part of residents' social and recreational lives.

The senior dining halls combine traditional dietary therapy with modern nutrition science, designing balanced and nutrient-rich menus according to seniors' physiological characteristics and health needs. Additionally, seniors can choose from various on-site themed restaurants, such as the food court, Yixin Garden, McDonald's, and Half-Hill Wellness Club, catering to different tastes and preferences (see Fig. 41).

**Fig. 37** Master plan of Gudian Wellness Park

**Fig. 38** "Colorful Yunnan · Gudian City" as an integrated cultural tourism and wellness complex

The park offers 869 finely decorated apartments, each 85 square meters, featuring two bedrooms, a living room, a dining room, a kitchen, and a bathroom, offering both spaciousness and privacy. Seniors can choose between Chinese- or Western-style interior designs. Each unit is fully equipped with furniture and appliances, including refrigerators, air

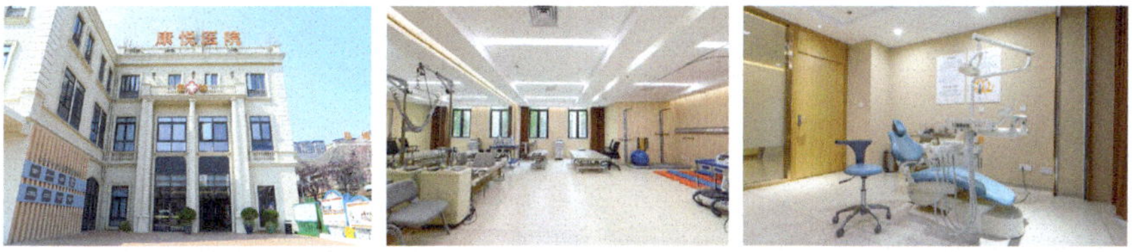


**Fig. 39** Dianchi Kangyue Hospital

**Fig. 40** Dianchi Senior University

**Fig. 41** Different restaurants in the health and wellness park

conditioners, televisions, washing machines, and internet access, ensuring a personalized and convenient living environment.

Public spaces are designed with seniors' needs in mind, featuring anti- slip floor tiles, wheelchair ramps, and handrails at entrances and corridors for safe mobility (see Fig. 42).

A professional health service center is staffed with trained caregivers providing comprehensive daily care, medical services, emergency call services, and 24/7 support, allowing seniors to fully enjoy their lives with peace of mind (see Fig. 43).

## 1.2 International Practices in Age-Friendly Service System Design

### 1.2.1 Japan–Kamakura Midori-tei

Inamuragaisaki, blessed with a temperate climate and proximity to ancient cultural heritage, has long been celebrated in ukiyo-e woodblock prints as a renowned scenic destination. The site offers panoramic views of the expansive ocean and the distant Mount Fuji .(see Fig. 44–45).

Strategically situated within this privileged setting, Kamakura-tei is a three-story architectural complex (see Fig. 46) that harmonizes with the region's rich seasonal landscapes (see Fig. 47).

Comprising 26 meticulously designed rooms, this premium-priced retirement residence embodies the accumulated design expertise of Benesse Style Care, a leader in elder care innovation (see Fig. 48).

GranInamuragasaki Kamakura Midori-tei is the first high-end fee—based retirement home in Kanagawa's business district operated by Benesse Style Care, with an entrance fee exceeding 20 million yen. To create untapped value, our company as the operator participated in the design process from the early planning stages, flexibly applying models and language to understand the lives of the end users—the residents.

Additionally, during the pre—opening phase, environmental colors and visual language were incorporated into the design scope to achieve a sense of unity between architecture, environment, and services. Their goal is not merely to stimulate consumption but to explore a reasonable form of elderly care facilities by treating the project as a "home" that grows more appealing with time facilities.

#### B5 Type Room (60.6 m$^2$)

A corner unit exceeding 60 square meters, featuring a bedroom separate from the living area, a

**Fig. 42** Design of age-friendly living spaces

private bathroom, and a compact kitchen. Equipped with underfloor heating and a nurse call system, this spacious layout is ideal for two occupants, offering ample freedom and comfort (see Fig. 49).

**A3 Type Room (30 m²)**
A third—floor ocean—view room designed for relaxation while enjoying the se aside scenery. The interior employs an accent cross motif in Shōnan-inspired color tones, complemented by abundant natural light for a soothing ambiance (see Fig. 50).

**B2 Type Room (40 m²)**
A self- contained unit with a modular bathroom and compact kitchen, optimized for solo living. Residents may personalize the space with their own furnishings (see Fig. 51).

Notably, the facility offers three specialized bathroom configurations tailored to elderly residents' needs: A hinoki (Japanese cypress) bathtub for aromatic wood immersion therapy. A barrier-free mechanical bath accommodating wheelchair users. A standard private bathroom .(see Fig. 52–53):

- A hinoki (Japanese cypress) bathtub for aromatic wood immersion therapy.
- A barrier-free mechanical bath accommodating wheelchair users.
- A standard private bathroom (see Fig. 53).

**Fig. 43** Professional nursing staff provide round-the-clock comprehensive care

**Fig. 44** Kamakura Tourmaline gate

### 1.2.2 Japan–Maihama Club

The arched doorways, columns, soft color palette, and expansive windows of Maihama Club evoke the grandeur of a European manor (see Fig. 54). This senior care facility was established by a Swedish entrepreneur who first immigrated to Japan in the 1990s. It operates as a premium retirement home, catering to those who prioritize quality and are willing to invest in it.

Opened in 2003, this high-end facility has since expanded to include dementia day-care services, short—term stays, a training center, and an on-site clinic (see Fig. 55). A variety of room types are available to accommodate different needs (see Figs. 56, 57, 58), and well-equipped communal spaces are provided to support residents' daily activities (see Figs. 59, 60).

**Fig. 45** Geographical location of Kamakura Tourmaline

**Fig. 46** Appearance of Kamakura Appearance of

The monthly fee of ¥575,000 covers accommodation, meals, and utilities. After deductions from Japan's national long-term care insurance, residents pay an additional ¥7,000 to ¥28,000 per month for support and nursing care, depending on their required care level.

**Six Key Features:**

Palliative Care Philosophy—Adopts the World Health Organization (WHO)-endorsed approach to comfort- focused care.

On-Site Cognitive Impairment Support—Implements specialized techniques and methodologies for dementia care.

Personalized Attention&Liaison System–Ensures individualized care through a well-staffed support network.

In -House Culinary Services—Provides high-quality, delicious meals via a dedicated food service team.

Strengthen cooperation and care with medical institutions through a 24- h care system.

**Fig. 47** Kamakura tourmaline architecture location landscape

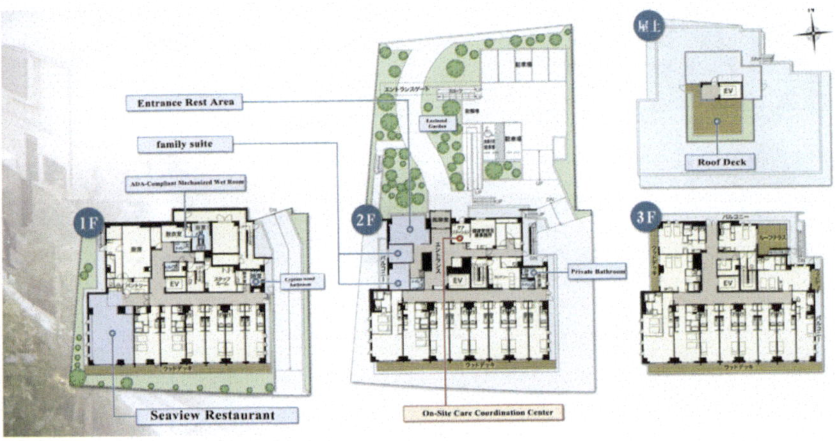


**Fig. 48** Kamakura tourmaline floor plan

### 1.2.3 United States–Parc Provence Senior Living Community

Parc Provence stands as a model for Alzheimer's and dementia care in the St. Louis region. This community specializes in assisted living and memory care, delivering exceptional support to seniors, particularly those with dementia and Alzheimer's, along with their families (Fig. 61).

As a vibrant community, Parc Provence enhances residents' quality of life by improving both physical well-being and cognitive function. Its staff is led by a team of seasoned, medically trained professionals—from administrators with nearly three decades of care experience to a medical director who holds a leadership role in geriatrics and dementia research at Washington University. Around-the clock, seven-day-a-week care ensures unparalleled support.

Every architectural detail, from interior spaces to landscaped grounds, is meticulously designed to foster warmth and privacy. The residence features an elegant Victorian-inspired design, comprising five distinct yet interconnected cottage-style buildings (see Fig. 62).

**Fig. 49** Kamakura tourmaline B5 room

**Fig. 50** Kamakura tourmaline A3 room

**Fig. 51** Kamakura tourmaline B2 type room

**Fig. 52** Kamakura tourmaline courtyard

**Fig. 53** Kamakura tourmaline matching bathroom

**Fig. 54** Architectural Appearance

Each unit functions as an intimate, home-like setting, complete with a refined dining area, living room, kitchen, and sunroom (see Fig. 63). Beyond the corridors lies a serene backyard garden, offering residents a secure sanctuary to connect with nature (see Fig. 64).

Parc Provence provides a secure and supportive living environment designed to make seniors feel truly at home, characterized by its "homelike" interior design that fosters familiarity and comfort.

Each residential unit provides access to a central activity area featuring a variety of engaging and inviting spaces, including a bistro, grand hall, art gallery, library, private dining room, and billiards lounge. The interior corridors are adorned with elegant fabrics, furnishings, artwork, and soft décor, transforming private and communal areas into serene and welcoming retreats. The "homelike" design incorporates bright and spacious kitchens, cozy living rooms, refined dining areas, private laundry facilities, charming rear porches, and beautifully landscaped courtyards. An advanced security system enables senior s to safely enjoy the lush gardens, patios, and meticulously designed communal spaces (see Figs. 65 and 66).

Each senior's suite is meticulously tailored to individual needs, accounting for varying levels of independence and care requirements. Within the warm and sophisticated ambiance of the "home like" suites, residents will find 3

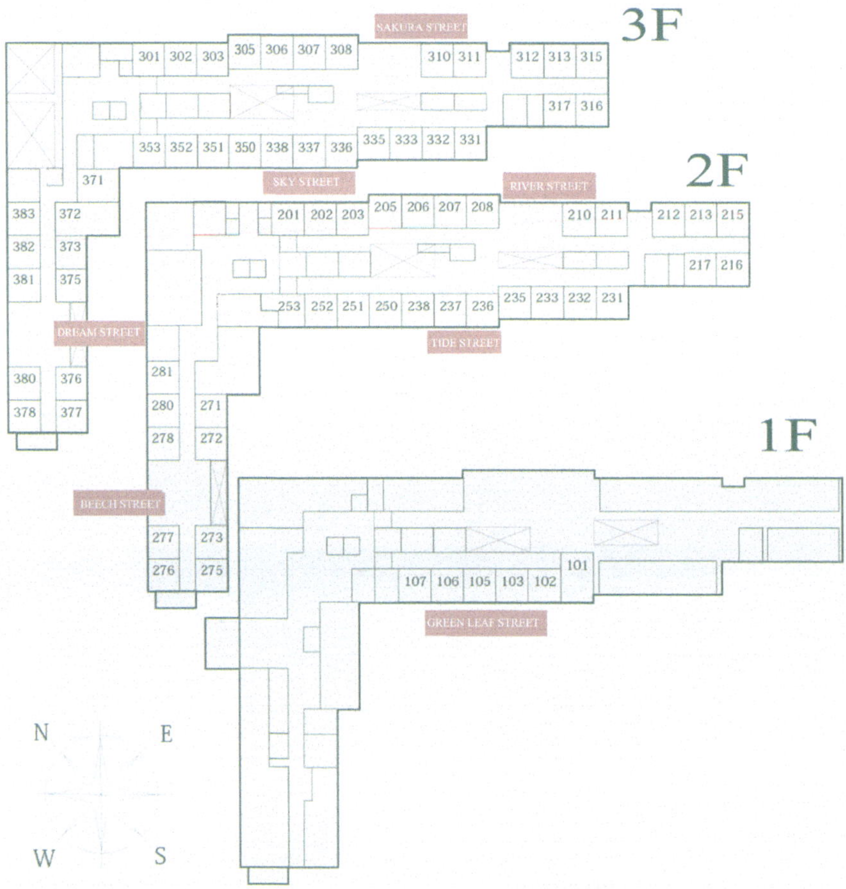


**Fig. 55** Plan of each floor of the building

-meter- high ceilings, bay windows with decorative moldings, coordinated bedding ensembles, elegant window treatments, plush carpets, and designer wallcoverings (see Fig. 67). Every suite also includes a 24 - h emergency response system. Whether seniors prefer a quaint, cozy atmosphere or an open, airy layout, Parc Provence offers a selection of floor plans to accommodate diverse personal preferences.

### 1.2.4 Sweden–Danvikshem Senior Care Facility

Danvikshem Senior Care Facility is located along the Baltic Sea coast. Constructed in 1914, its castle-like architecture was originally designed as a hospital (see Fig. 68–70). As early as 1551, Gustav Vasa, the founding king of Sweden, issued a decree to eradicate poverty and disease in Stockholm. He donated a portion of his personal wealth to establish Sweden's first national nonprofit foundation, which funded the construction of this hospital—now the Danvikshem Senior Care Facility.

For centuries, the foundation's mission has been to provide compassionate care, nourishment, and shelter to those in need, a principle that remains central to the facility's operations. The care philosophy at Danvikshem is built upon person—centered care and team-based service delivery.

Person—Centered Care: This approach emphasizes respect for the individual as a whole, addressing not only physical ailments but also psychological challenges such as anxiety,

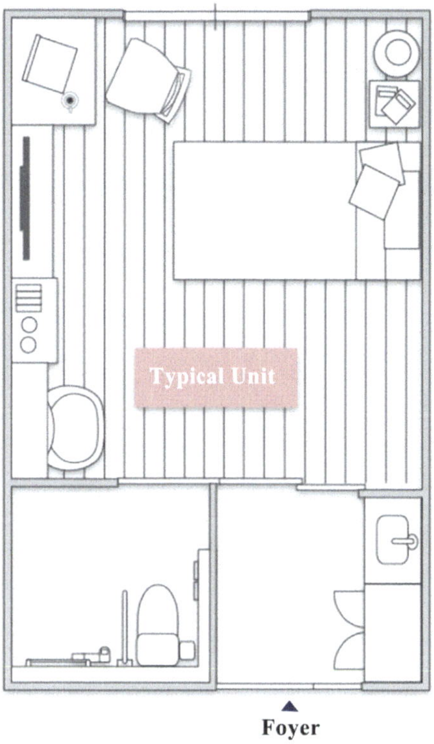


**Fig. 56** Standard room floor plan

depression, and agitation. It ensures that residents and their families are actively involved in medical decision-making and care planning (Figs. 69 and 70).

Team-Based Service: Care is delivered through a multidisciplinary team comprising caregivers, nurses, rehabilitation specialists, physiotherapists, and physicians. Beyond resident care, institutional guidelines mandate a safe and supportive work environment, encouraging staff participation in workplace improvements.

Currently, the facility accommodates 206 seniors, supported by a workforce of over 220 employees.

In addition to basic medical and nursing services, the facility employs licensed physiotherapists who oversee daily rehabilitation exercises for residents and contribute to designing group exercise programs. A certified dental practitioner is also available, providing comprehensive oral healthcare services (excluding dental implants and cosmetic procedures).

The facility features a 300-seat chapel, which regularly hosts musical performances (see Fig. 71). Beyond its renowned person-centered

**Fig. 57** Standard twin room—actual view

**Fig. 58** Standard single room—actual view

**Fig. 59** Public areas

**Fig. 60** The drawing room

care, Danvikshem boasts well-equipped ancillary facilities (see Fig. 72).

Furthermore, the facility has established two specialized the-rapeutic spaces.

**Reminiscence Room**

Designed to evoke memories of the past, this space aids in cognitive therapy for residents with dementia while facilitating meaningful interactions between caregivers, family members, and residents. The room contains period-appropriate furnishings, toys, toolkits, washing machines, and other 20th -century artifacts. A distinctive feature is the Google Earth-connected stationary bicycle, allowing residents to virtually cycle through global destinations without leaving the facility (see Fig. 73).

**Sensory Stimulation Room**

This therapeutic environment employs multi sensory activation through specialized equipment including:

Music-integrated beds
Dynamic lighting systems
Tactile-stimulation armchairs (see Fig. 74)

**Fig. 61** Architectural appearance

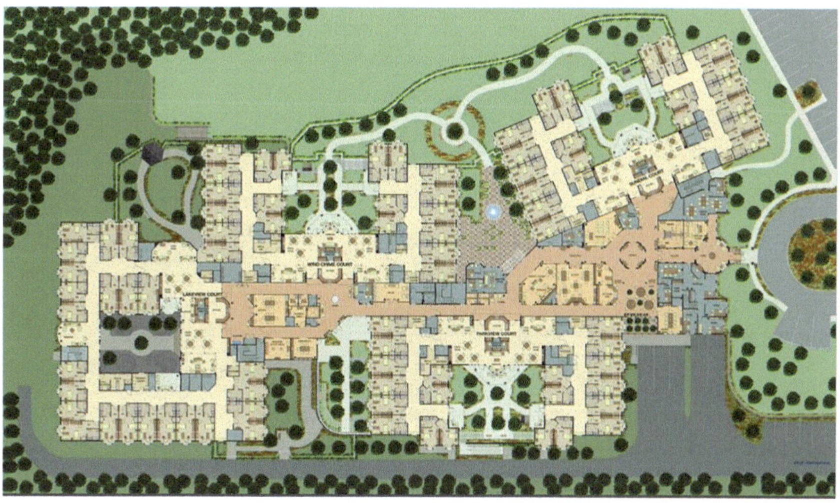

**Fig. 62** Overall plan of nursing home park

### 1.2.5 Singapore–Kampung Admiralty Integrated Senior Living Complex

Kampung Admiralty Community Complex (also known as the Admiralty Village Project) is a public housing development by Singapore's Housing & Development Board (HDB), integrating public facilities and services into a single mixed—use complex. Designed to address Singapore's aging population, the project is exclusively for residents aged 55 and above.

The Kampung Admiralty Complex consolidates a wide range of public amenities within a single building. The architects explain: "The conventional approach would allocate separate plots of land to different government agencies, resulting in multiple standalone structures. In contrast, this one-stop integrated complex

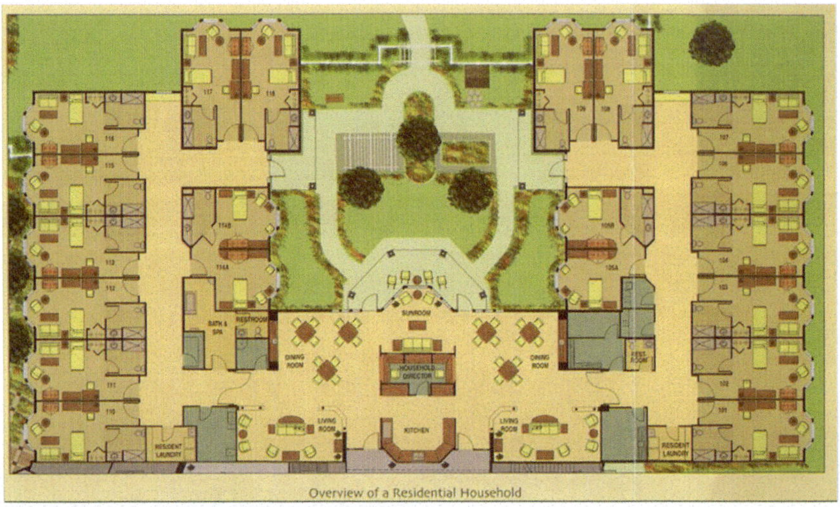


**Fig. 63** Plan of a standalone building in a nursing home

**Fig. 64** Outdoor landscape

maximizes land efficiency, serving as a model for meeting the needs of Singapore's aging society."

Due to site constraints and a height restriction of 45 m (148 feet), WOHA Architects designed a vertically layered 'village' composed of three distinct strata (see Fig. 75). The complex is conceptualized as a "club sandwich".

The lower stratum accommodates a community plaza (see Fig. 76),
The middle stratum houses a medical center,
The upper stratum features a community park with senior housing

The design team emphasizes that "the close integration of healt-hcare, social, commercial, and convenience facilities foster s intergenerational bonding and promotes active aging."

The three-tiered spatial configuration incorporates diverse architectural functions, fostering cross- functionality and programmatic diversity while reserving ground—level spaces for public activities. This design philosophy minimizes

**Fig. 65** Indoor public areas

**Fig. 66** Indoor public areas

**Fig. 67** Indoor scene of suite

**Fig. 68** Architectural appearance

**Fig. 69** Landscape of nursing home courtyard

**Fig. 70** Location natural landscape

**Fig. 71** Church

**Fig. 72** Supporting facilities in public areas

**Fig. 73** Reminiscence room

**Fig. 74** Sensory stimulation room

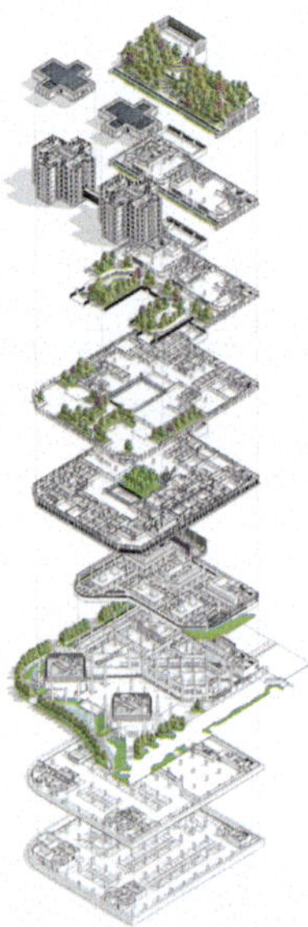

**Fig. 75** Architectural exploded axonometric diagram

distances between healthcare, social, recreational, commercial, and convenience facilities, strengthening intergenerational connections and advancing active aging (see Fig. 77).

The upper–level community park (see Fig. 78) features terraced greenery, an open community farm, and a social hub equipped with benches and fitness stations, creating an inclusive gathering space for seniors (see Fig. 79). Within this warm, welcoming environment, the public can participate in organized activities, seasonal festivals, shopping, or dining at the hawker-style food court on the second floor (see Fig. 80).

The inclusion of an on-site medical center eliminates the need for residents to travel to hospitals or undergo day surgeries due to logistical constraints. To enhance healthcare accessibility, the clinic's consultation and waiting

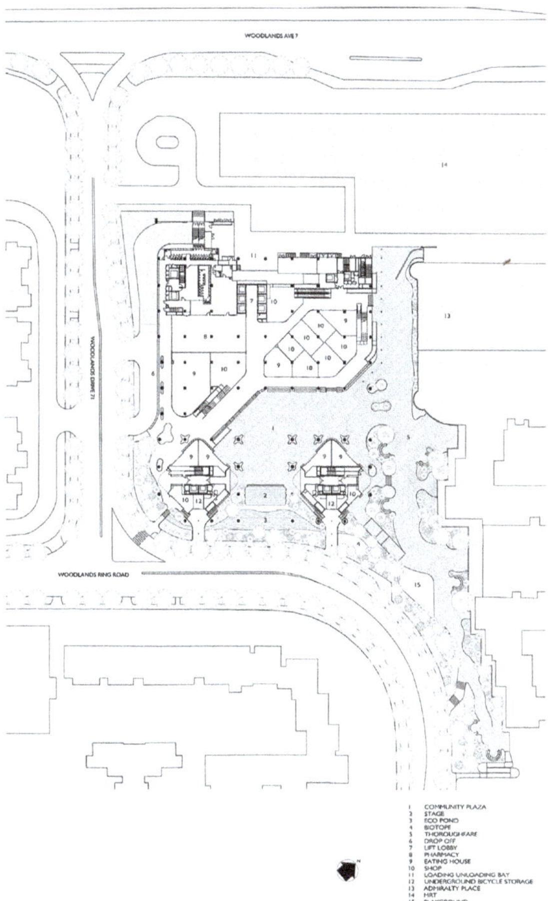


**Fig. 76** First floor plan of the building

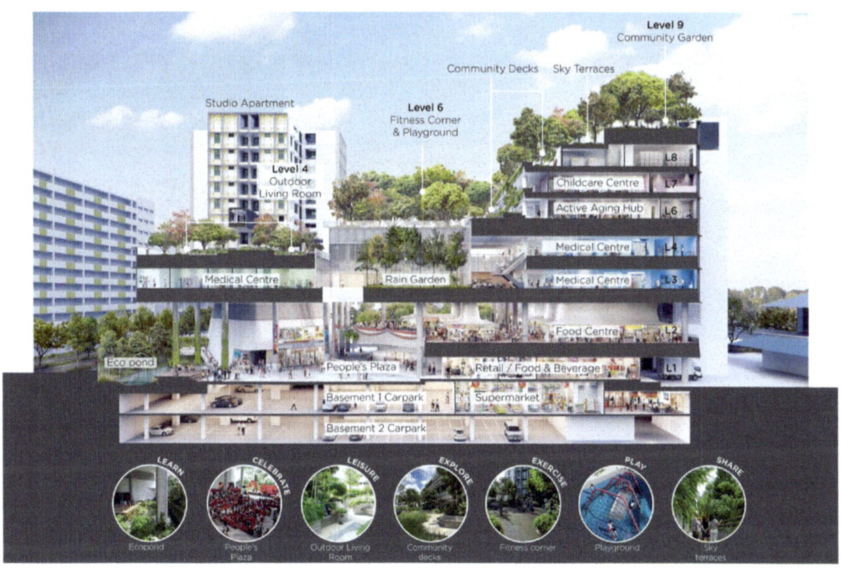


**Fig. 77** Architectural section drawing

**Fig. 78** Architectural bird's-eye view

**Fig. 79** Rooftop park

**Fig. 80** Public plaza

area s utilize natural lighting through perimeter windows and a central atrium, ensuring a comfortable patient experience. The medical center is arranged around a central courtyard, offering soothing green views (see Fig. 81).

# 2 Definition and Significance of Age-Friendly Services

## 2.1 Vision and Future Trends of an Age-Friendly Service System Promoting Social Inclusion

With the accelerating global aging population, establishing an age -friendly service system that upholds the dignity and value of older adults while seamlessly integrating modern technology and social care has become a critical mission of our era. This system not only aims to meet the growing demands of senior s in healthcare, emotional support, and social engagement, but a l so emphasizes breaking age barriers through intelligent, personalized service de-sign, fostering intergenerational communication and mutual understanding to achieve genuine social inclusion.

In the future, age-friendly service systems will deeply integrate cuttingedge technologies such as big data, artificial intelligence (AI), and the Internet of Things (IoT) to provide seniors with more precise and convenient service experiences. From smart wearable s monitoring health conditions to telemedicine reducing healthcare access barriers; from smart home systems enhancing daily convenience to community service platforms facilitating neighborhood mutual aid and social activities— each technological application will profoundly transform seniors' lifestyles, making their later years more vibrant, secure, and comfortable.

Moreover, we anticipate broader participation from policymakers, businesses, social organizations, and the public to collectively drive the continuous improvement and innovation of age-friendly service frameworks. By strengthening cross- sector collaboration, we can explore diverse service models, such as "time bank" mutual support mechanisms and "silver economy" strategies that encourage senior s to contribute to socioeconomic development. These approaches will further unlock the social vitality and creativity of older adults, enabling them to be active contributors to society while benefiting from supportive services.

Ultimately, our pursuit extends beyond technological advancements to encompass a profound cultural shift, where respect, care, and support for the elderly become a shared societal value and practice. Together, we can envision a harmonious, inclusive future where seniors enjoy fulfillment, security, and joy in their later years. This vision represents not only a commitment to the well-being of older generations, but also a strategic foundation for sustainable societal development.

**Fig. 81** Health and wellness hub

# References

1. Zuo Meiyun. Smart Elderly Care: Service and Operation[M]. Tsinghua University Press, 2022.
2. Zhou Yanmin. Elderly Housing[M]. Second Edition. Beijing: China Architecture & Building Press, 2018.
3. Luo Shijian, Zou Wenyin. Current Status and Progress of Service Design Research[J]. Packaging Engineering, 2018, 39(24): 43–53.
4. Jiang Songrong. "The Fourth Principle"—Design Ethics Research[J]. Ethics Research, 2009(02): 57–62.
5. Wu Ping, Peng Yali. Innovation Design for Aging [M]. Beijing: Chemical Industry Press, 2022.
6. Gao Yunpeng, Hu Junsheng, Xiao Jian, Lao Psychology[M]. Beijing: Peking University Press, 2013: 52–128.
7. Liu Na, Cao Shengsheng. Research on the fault-tolerant design path of somatosensory interaction under the background of aging[J]. Design, 2024, 37(08): 130–133.
8. Wang Meng, Hu Yongsheng, Guo Hong. Analysis of indoor space environment for the elderly based on emotional needs[J]. Design, 2022, 35(11): 79–81.
9. [US] Donald Norman. Design Psychology[M]. translated by Mei Qiong, Beijing: CITIC Press, 2010.
10. Victor Rainier. Residential environment design in the aging era - innovative practice of assisted living facilities[M]. Translated by Qin Ling et al. China Architecture & Building Press, 2019.
11. Tang Wen, Yu Yun. Research on digital humanities innovation and transformation design of elderly education[J]. Aging Research, 2024, 11(5): 2123–2137.
12. Li Guangdong, Research on interactive design of mobile-friendly interface for the elderly[J]. Journal of Jilin Academy of Arts, 2024(01): 58–63.
13. Zhou Yanmin, Li Jiajing. Research on therapeutic space environment design of nursing institutions for demented elderly people[J]. Journal of Architecture, 2018(2): 67–73.


W. Tang, *Aging-friendly Service System Design*, https://doi.org/10.1007/978-981-95-2054-1

14. Li Zhiyu, Xu Jiayi, Nie Qian. Research on the interface design of the medical companion app for the elderly[J] . Chinese Art, 2023(01): 72--78.
15. Sui Yong. Visual design of Internet products (Web\Mobile Web\APP) (style)[M]. Beijing: Tsinghua University Press, 2015.
16. Wang Jie. Construction of an aging-friendly support system for the digital society[M]. Beijing: Publishing House of Electronics Industry, 2023.
17. [US] Amos Rapp. The meaning of the built environment: non-verbal expression methods[M] . Translated by Huang Langu. Beijing: China Architecture & Building Press, 2003: 41.
18. China Academy of Information and Communications Technology, Internet Society of China. Digital Benefits for the People: A Collection of Excellent Case Studies on Internet Applications for the Elderly and Accessible Practices[M]. Beijing: People's Posts and Telecommunications Press, 2023.
19. Tang Wen, Zhang Yanyan. "Longevity" Design—A New Form of Intelligent Elderly-friendly Space Design[J]. Design Art Research, 2023, 13(04): 93–97.
20. Wang Lijun, Liu Rongling, Sun Xuyang. Research on Emotional Design and Aging-friendly Urban Micro-public Spaces[M]. Wuhan: Huazhong University of Science and Technology Press, 2023.
21. Zhang Yanyan, Gao Zhitong, Tang Wen. Research on the Spatial Renovation Design of Nursing Homes Based on the Emotional Needs of the Elderly—Taking the Spatial Renovation Design of Nursing Homes in Wuhan as an Example[J]. Design, 2023, 36(15): 46–49.
22. Ouyang Hongbin. Configuration of Nursing Homes under the Background of Aging - Based on the Theory of Resilient Cities[M]. China Architecture & Building Press, 2021.
23. Liang Yuqi, Tang Wen. Value, theory, and system: dual-carbon elderly care in China's modernization process[J]. Aging Research, 2024. 11(5): 2001–2010.
24. Richard Weston. Materials, forms and architecture[M]. Translated by Fan Suning and Chen Jialiang, Beijing: China Water Resources and Hydropower Press, 2005.
25. Ivy Friedman. Adaptive housing[M] . Translated by Zhao Ying and Huang Qian, Jiangsu: Jiangsu Science and Technology Press, 2004.
26. Wei Dake, Duan Xingkuai. Research on the path finding performance of traffic space in elderly care facilities based on VR experiment[J]. Journal of Architecture, 2023(S2): 88–93.
27. Zhou Yanmin. The needs of the elderly for room functional layout[N]. China Real Estate News, 2013
28. Ding Yingshun. Japan's aging population and elderly human

resource development[M]. China Social Sciences Press, 2016.
29. Smart City Editorial Department. The State Council issued the "13th Five-Year Plan" National Aging Development Plan[J]. Shanghai Collective Economy, 2017.
30. [UK] Robert Feldman. Developmental Psychology: Lifelong Development of Human Beings 4th Edition[M]. translated by Su Yanjie, World Book Publishing House, 2007.

The manufacturer's authorised representative in the EU is Springer Nature Customer Service Centre GmbH, Europaplatz 3, 69115 Heidelberg, Germany. If you have any concerns regarding our products, please contact ProductSafety@springernature.com

Printed and bound by CPI Group (UK) Ltd, Croydon, CR0 4YY
25/09/2026
02223058-0001